THE **PLANT LOVER'S GUIDE** TO
SEDUMS

BRENT HORVATH

TIMBER PRESS
PORTLAND · LONDON

CONTENTS

WHY I LOVE SEDUMS

I grew up in the nursery business and continue to work in it today, and throughout my career, I have come to admire sedums more and more. **They are so easy to grow, so varied, and so gorgeous.**

In autumn, the rich, earthy colors of the season—russets, golds, and buffs—signal cooling temperatures and the winding down of the gardening year. Many people grow Autumn Joy *Hylotelephium* 'Herbstfreude', the classic perennial sedum, and enjoy its season-long metamorphosis of hue. The flowerheads of soft raspberry pink deepen to garnet as fall approaches. What's more, in all but the harshest climates, this tenacious plant continues to delight through the winter months as its flowerheads turn copper, then bronze.

Something of a sedum poster child, 'Herbstfreude' may be the most familiar but is by no means the only worthy example of this broad group. It has many charming cousins.

Sedums make perfect container plants. Even if you don't have a garden, you can still enjoy them.

Of the literally hundreds of different sedums, tall and small, the majority are as easy to grow as 'Herbstfreude', asking little more than a sunny spot in well-drained soil. Because sedums are technically succulents, they can survive drought and forgive neglect, thanks to their juicy, moisture-retaining leaves and tough root systems. Unlike their relatives the jade plant (*Crassula*) and cacti, many sedums are hardy in USDA Zones 5 to 9. If you live in a region where sedums cannot withstand winter outdoors, you can always enjoy them in containers and bring them indoors for the cold months, or take cuttings and start over the following spring.

If you don't have any sedums in your garden yet, start small. Pick one or more for a pot or for your garden. I highly recommend Christmas cheer (*Sedum ×rubrotinctum*) for a container. This trusty plant will brighten a window for the winter, cheerfully color up in spring, enjoy the heat of summer, and hunker down with you in the fall.

Getting to know sedums has been an adventure for me. There are so many species, from all over the world. Some hail from the steppes and lower mountain slopes of eastern Asia and Europe, others come to us from settings as diverse as Japanese islands and Rocky Mountain outcroppings. Because gardeners have appreciated these plants for a long time, there are plenty of worthwhile selections, hybrids, and variations, and new ones are continuing to enter the trade every year.

The last definitive book on the group, *Sedum: Cultivated Stonecrops*, by Ray Stephenson, was published in 1994. A lot has changed since then. Not only have many more plants and variations come on the scene, but botanical research has entered a new realm whereby plants are now being classified and reclassified according to DNA-based information. As botanists have been working to sort out and catalog the diversity in sedums, there has been a lot of reorganization and some confusion persists. Briefly, many stonecrops remain in the genus *Sedum*, although some have been reclassified into other genera, notably *Petrosedum* and *Phedimus*. Most if not all of the border types are now placed in a separate genus *Hylotelephium*. Please don't let such details daunt you or deter you from exploring these plants or tracking down ones that you really want. No matter what they're called, they are all related and often marvelous.

Whether you favor the handsome border sedums with their surprising variety of flower and foliage colors, or you prefer the low-growing groundcover sedums, there are plenty of plants to capture your imagination and enhance your garden—not just in autumn, but throughout the gardening year.

Border Beauties

The taller, upright-growing sedums, now in genus *Hylotelephium*, grow more or less erect and have dense, domed flowerheads composed of many tiny starry flowers, much like a colorful broccoli head. Foliage is thick, fleshy, and succulent and runs from sage green to blue to dark red in color.

Most border sedums come to life in spring with light green leafy buds that eventually open to white, pink, or red flowers in late summer to early fall. In summer, they are a popular stage for butterflies (avid butterfly gardeners include sedums on their lists of desirable plants, particularly the autumn-blooming border types), and bees often visit as well. As the weather cools, the flowerhead colors grow richer and deeper. If your winters are not too harsh, these sedums remain standing for months longer still—their broad forms contrasting with the faded foliage of nearby plants or perhaps gathering a jaunty cap of snow atop their dried flowerheads.

Some plants are tall, up to 2 feet (60 cm), and can hold their own in a mixed border. Recent years have seen the introduction of some intriguing shorter ones. The so-called cloud sedums are not only a foot (30 cm) tall and stockier, but also extra-floriferous as well. The Desert series, which includes 'Black', 'Blonde', and 'Red', is even shorter, coming in under a foot (30 cm) tall. These introductions suggest new uses.

In general, the border sedums thrive in the company of other plants because they are undemanding and not aggressive. Many gardeners include them among late-season bloomers like asters, ornamental goldenrods, and boltonias, where their broad flowerheads contribute harmonizing or contrasting color. They're terrific with ornamental grasses, which also come into their own in the fall.

There are hundreds of *Hylotelephium* cultivars and hybrids, and more coming out each year. In this book, I have included about two dozen of the ones I think are the best or most representative.

Groundcovering Sedums

These little charmers, collectively known as stonecrops, usually remain under 4–6 inches (10–15 cm) tall as they creep, spread, and sprawl. You might be surprised to learn that they are so closely related to the border types, given their small stature, but in many respects, they are quite similar. They have succulent leaves (which tend to be evergreen in mild-climate areas), usually carry their starry flowers in clusters, and for the most part thrive in average soil and full sun.

Because of their size, stonecrops are often used as botanical carpets, on either open ground or lounging over rocks and walls. A few are tough enough to withstand foot traffic

Few plants can compete with sedums for sheer range of colors and versatility, not to mention beauty.

A botanical carpet of *Sedum mexicanum* 'Lemon Ball' defines an area and flatters focal-point plants in a formal garden.

and adorn the spaces between terrace or walkway stones. Some are so tiny, they can enhance a terrarium display, and bonsai enthusiasts use them at the base of their treasured miniature trees. Individual plants, especially those with striking foliage or flowers, are ideal in rock gardens. These sedums can also soften the edges of formal displays in urn-type planters, trough gardens, and windowboxes; solo, they'll thrive in shallow pots filled with a gritty or well-drained soil mix. They're also excellent for greenroofs and other innovative planting concepts.

Many gardeners prefer stonecrops by themselves or paired with one another. Thanks to the different hues these come in, and their changeable nature over the course of a growing year, stonecrops are indeed wonderful combiners. Their color show can be enjoyed in a mass planting, or up close and intimate in a potted display. Aim for contrast. Be daring, have fun.

Last but not least, these sedums are excellent with other succulents, such as hens and chicks, because they thrive in the same settings. The slightly larger scale of such companions makes for dramatic contrast.

If you become enthusiastic about sedums, your garden will grow. You will find yourself composing interesting and colorful new displays and experimenting with new varieties. You will start seeking out intriguing and sometimes rare, new additions.

While this book covers many different sedums, there are a few others I'd like to mention. Some are just unusual, some are rare but becoming available, and some are brand new. For instance, I've been able to track down seed of the annual *Sedum caeruleum*, which, while not common, sometimes is available. It comes from the Mediterranean, so I plan to grow it in sandy soil. It has blue flowers and, at times, red foliage. Not only is it the only blue-flowered sedum, it is the only blue-flowered member of the crassula family (Crassulaceae). It is beautiful.

Another plant I've been thrilled to acquire is *Sedum booleanum*, which has silver foliage and orange-red flowers. What a hot combination! Any of the North American natives are of interest to me, but many are still rare in the trade. *Sedum obtusatum*, though not common, can be found and is worth the search. Others are species that I covet for breeding purposes, like *Hylotelephium pallescens*, which has thinner gray-green foliage on upright plants. My point is, there is still much more to be found, learned, and enjoyed in the world of sedums.

In short, sedums are a dynamic group of plants at this time. Not only are the names constantly being revised and updated, but new introductions are appearing on the market every year and new species continue to be brought into cultivation. Do your best to find out the correct name of each one you bring home and certainly to accompany any plant or cutting you share with someone else, but don't let concern for correctness bog you down too much. There are just too many tempting discoveries and possibilities. You will not find a more easy-going, rewarding group of plants. I hope the information in this book will empower and inspire you to include more sedums in your garden and potted displays. Come on in, the wonderful world of sedums awaits you.

DESIGNING WITH BORDER BEAUTIES AND STONECROPS

The great diversity of sedums in form, foliage, and flower allows a lot of latitude and creativity for gardeners. Depending on where you live and the nature of the site, the ways of using these plants are as endless as your imagining. No matter how you ultimately choose to use the many different sedums, bear in mind a few basic design principles:

Scale and Proportion. The smaller the space, the smaller the sedum needs to be to be appropriate. Also the larger the space, the larger a planting should be.

Repetition. This is an area where sedums can excel, because they are easy to plug in here and there, thus creating a repeating texture or color throughout a flower border or garden. It helps, too, that sedums increase easily, either with your intervention or due to their own natural spreading tendencies.

Contrast. This is another easy role for sedums to fill because they come in almost any color or texture you might be seeking. Depending on the surrounding plants, a boost of emphasis can be achieved with sedums. They can be the added ingredient that flatters or sets off a focal-point plant, highlighting its features and helping it to stand out better.

Texture and Form. Sedums offer a multitude of choices—from fine to rough, slender or needlelike to chubby, smooth and glossy to glaucous (having a gray, bluish, or whitish

Hylotelephium sieboldii repeats the arching stem of the evergreen while contrasting strongly with its color and texture.

coating on the leaves that is easily rubbed off)—to set off most anything. Their presence not only looks interesting and attractive juxtaposed with many other plants, but also against nonplant parts of a garden like a walkway, a patio, rocks and stones, or other hard solid surfaces.

Now let's look at some of the more common or popular uses first. To oversimplify, sedums have two primary uses in the garden. The first is as a border plant; the second is as a groundcover. Some sedums may be used both ways.

Border Sedums

The first major use of sedums is as border plants. Here we are talking about the larger, upright-growing perennial types. These are in their glory in the fall, which is why they are sometimes called "autumn sedums." Botanists have now set them apart in their own genus, *Hylotelephium*. The most familiar and popular one, of course, is Autumn Joy *Hylotelephium* 'Herbstfreude'. This resilient plant has true four-season appeal, emerging in spring as a blue-green dome with toothed foliage, developing light green broccoli-like buds in mid to late summer, and turning pink to red in fall and finally to cinnamon brown as the flower fades to a dried stem by winter.

All hylotelephiums follow this same cycle, though foliage and flower colors vary. You can make a gorgeous fall border "vignette"—as landscape designers term simple, effective plant combinations—with border sedums and other perennials that are at their best at that time of year. The rich shades of the flowerheads and the often-colorful leaves blend very well with plants of red, yellow, or purple flowers or foliage. Alternatively, try a sedums-only display, mixing a few closely related species or cultivars for a blend of colors. This tack results in a handsome, low-maintenance ribbon of color, or brings a multihued, tapestry-like look to a larger or broader bed.

There are many outstanding choices today and more coming out all the time, all of which create beautiful displays. The following are a few of my favorite combinations, which you can follow or use as inspiration for your own compositions.

A classic fall combination of *Hylotelephium* 'Herbstfreude', *Pennisetum alopecuroides* 'Red Head', and *Perovskia atriplici-folia* 'Superba'. Thanks to their attractiveness, dependability, and varied hues, border sedums are an excellent choice for mixed borders.

A mass planting of *Hylotele-phium* 'Matrona' along a walkway at the Morton Arboretum in Lisle, Illinois. Notice how the colors blend beautifully with autumn hues in the surrounding landscape.

By combining similar sedums, you can extend the season of interest. Ornamental grasses are always a nice backdrop; shown here are *Calamagrostis acutiflora* 'Karl Foerster' and 'Overdam'. Combining species and varieties can also keep disease from spreading.

WITH CONEFLOWERS, ORNAMENTAL GRASSES,
AND OTHER PERENNIALS

The native American wildflower purple coneflower (*Echinacea purpurea*) has seen a renaissance in recent years. Nurseries are now offering these durable, easy-going perennials in all sorts of wonderful colors beyond the traditional pink-purple, including orange, magenta, yellow, and white. They like the same growing conditions as border sedums—decent, well-drained soil in full sun—and bloom well into autumn, so they make perfect companions.

I find that border sedums fit right in with some of the new, more dwarf coneflowers. I love pink *Hylotelephium* 'Red Cauli' or *Hylotelephium spectabile* 'Neon' with short, pink *Echinacea* 'Pixie Meadowbrite' or 'Pow Wow Wild Berry'. For a different, softer look, grow a white form like *Hylotelephium* 'Thundercloud' with a dwarf white coneflower like *Echinacea* 'Snow Cone' or 'Baby White Swan'.

Ornamental grasses, particularly the shorter ones, are also natural companions. Their thin leaves contrast well against the rounded, fleshy *Hylotelephium* foliage. If you choose one with colorful blades and/or seedheads (plumes), particularly ones that change color in autumn, you'll have a winning picture.

Combinations I've seen and like are gold-leaved grasses in the company of a red-flowered or dark red-leaved form like *Hylotelephium telephium* 'Purple Emperor' or *H.* 'Black Beauty'. Some good choices are *Hakonechloa macra*, especially the gold-leaved form 'Aureola', gold-leaved *Acorus gramineus* 'Ogon', and the grasslike sedge *Carex elata* 'Aurea'.

Because so many of these border sedums have some blue to their foliage, partnering them with other blue-hued plants makes an appealing display. The blue need not be a perfect match. In fact, contrast and echo are desirable. By the way, once you have a duo in place, you'll discover that the blues blend easily with additional colors, should you wish to expand your display.

Here, then, are some examples. Good blue ones include *Hylotelephium sieboldii*, *H.* 'Pure Joy', *Hylotelephium telephium* subsp. *ruprechtii* 'Hab Gray', and *H.* 'Thundercloud'. Combine these with short blue grasses like blue fescue (*Festuca ovina*) cultivars, blue oat grass (*Helictotrichon sempervirens*), or little bluestem (*Schizachyrium scoparium*). Whether a small grouping or a massed display with some or all of these plants, you are sure to enjoy an unusual and gorgeous sight, especially in the autumn months.

If you'd like to mix your border sedums with red-hued grasses, there are many good options. *Pennisetum setaceum* 'Rubrum', a fountain grass selection with red-purple foliage, always adds excitement, or introduce a grass with red-colored plumes, such as *P. alopecuroides* 'Red Head' or a smaller form like 'Ginger Love'. Once the border sedums are finished blooming and their seedheads turn brown, cinnamon-red-leaved plants like *Carex buchananii* or *Phormium tenax* add repetition to the color but sharply contrast in foliage.

Other ornamental grasses that combine well with the upright-growing (border) sedums include species of *Calamagrostis*, *Deschampsia*, *Eragrostis*, and *Molinia*. All are fairly low maintenance and like the same growing conditions that sedums share.

> Ornamental grasses, particularly the shorter ones, are also natural companions.

Hylotelephium telephium 'Red Cauli' with autumn moor grass (*Sesleria autumnalis*).

Perennials to Combine with Border Sedums

Recommended hardy companion plants:	Recommended tender companion plants:
Achillea	*Aeonium*
Agastache	*Aloe*
Allium	*Echeveria*
Geranium	*Graptopetalum*
Heuchera	*Pachyphytum*
Lavandula	*Senecio*
Nepeta	
Penstemon	
Symphyotrichum (Aster)	

The silver foliage of woolly speedwell, *Veronica incana* 'Pure Silver', along with *Sedum sexangulare* 'Golddigger', at the edge of a gravel garden.

Late summer into autumn offers grand opportunities for dramatic perennial-border displays that include *Hylotelephium* choices, but also some of the lower-growing, groundcovering ones discussed earlier in this chapter. Begin with a red- or purple-leaved *Hylotelephium*, or a sprawler like *Phedimus spurius* 'Fuldaglut' (fireglow). Add some red or orange flowers, such as an orange daylily (*Hemerocallis* 'Primal Scream') or butter-fly weed (*Asclepias tuberosa*). Try *Hylotelephium telephium* 'Cherry Truffle' with the hot pink flowers of *Persicaria amplexicaule* 'Firetail'. A simple yet daring combination would be any one of these hot-colored perennials underplanted with the rare orange foliage of *Sedum adolphii*.

If you feel the need to cool things down, try the red-leaved *Hylotelephium* types with silver foliage plants. Two fine choices are lamb's ears (*Stachys byzantina*) and *Artemisia schmidtiana* 'Silver Mound'. You could throw in some soft yellow yarrow, such as *Achillea* 'Moonshine' with its lacy, silvered foliage, to lower the temperature even more.

Purple flowers also go well. The purple spikes of *Liatris spicata* will help keep things cool in a border of colorful *Hylotelephium* plants. Fall favorites, the fall asters are another good option, for instance, the dense, purple-studded mounds of *Symphyotrichum (Aster) novae-angliae* 'Purple Dome'. Consider *Kalimeris incisa* 'Blue Star', an aster relative that reblooms in the fall.

Blue perennials are yet another option. A dependable perennial that combines well with all types of sedums is *Agastache*. It now comes in a wide variety of colors and sizes; the best blue is probably 'Blue Fortune', a long-blooming variety. *Eryngium planum*, which has mostly white or blue flowers and silver foliage, is another fine companion.

A trouble-free ribbon of *Hylotelephium* 'Herbstfreude' at a cemetery.

Hylotelephium 'Herbstfreude' used as a hedge along a lawn at the Boerner Botanical Gardens, Milwaukee, Wisconsin.

AS HEDGING

You can use border sedums by themselves to make an unorthodox but successful and attractive deciduous hedge in your home landscape. Just select many individuals of the same variety and plant closely. If your goal is a more formal look, some of the newer, shorter, heavy-flowering varieties are your best choice because of their size and strict habit. Examples include *Hylotelephium* 'Beach Party' and *H*. 'Thundercloud'.

WITH OTHER SUCCULENTS

Because they tend to look similar and have similar cultural requirements, many sedum cousins make great companions, in the garden as well as in containers. I encourage you to do some exploring and experimenting. The family sedums belong to, Crassulaceae, has many options, over 35 genera. Here are a few suggestions to get you started.

Hens and chicks (*Sempervivum*) are rosette-formers with often-colorful, starry flowers carried on spikes. They tend to be on the smaller side, but gain impact when grouped or massed.

Beard of Jupiter (*Jovibarba*) plants are quite similar to hens and chicks, except their "chicks" form on brittle stalks. Also their tiny flowers are bell-shaped and tend to face outward or down.

Echeveria species and cultivars have fleshy rosettes that resemble large hens and chicks in eye-catching shades of purple to blue. Echeverias sport succulent flowers of sunset-colored pink or yellow.

Kalanchoe species and cultivars are popular as houseplants and gift plants. These may also be enjoyed with sedums in pots or in the garden. Their succulent foliage is mostly green or silvered with red highlights, but they generate plentiful brightly colored flowers.

Although jade plants (*Crassula*) can grow quite large in a big container or outdoors in mild climates, they still make fine companions for different sedums, including the bigger, border-type ones. Their flowers tend to be in the less-dramatic white to pink range, so it's the contrast of foliage you'll want to highlight instead.

Groundcover, or Stonecrop, Sedums

The second major use of sedums is as groundcovers. Groundcovers are often important in our landscapes in spots where gardeners are seeking alternatives or need solutions to challenging settings. Many of the stonecrop-type sedums grow and prosper where grass cannot, because the soil is lean or gravely or because you don't want to invest in or fuss with a lot of watering and fertilizing and mowing. They can take the place of a lawn in difficult situations like a slope where you can't mow or in extra-dry, poor, rocky soil areas where you simply can't irrigate. In such spots, they will also act to prevent erosion.

Stonecrop-type sedums can be a good alternative to other groundcovers, too. They are more manageable than a patch or embankment of rampant ivy or goutweed. They can grow thickly, naturally keeping weeds at bay. They don't require much maintenance—you won't be spending time constantly clipping them back or ripping out parts that have gotten out of bounds.

They are good groundcovers for aesthetic reasons, too. They are good-looking plants in and out of bloom. Many offer three seasons of changing interest: greener foliage in spring and summer, foliage combined with a flurry of flowers in summer, and glorious fall color in both their leaves and their flowers. Others are fully evergreen, a plus in areas where winters are milder and the garden can remain attractive during those cooler months. The texture of their foliage and, oftentimes their colors as well, can create continuity.

Their function in an overall plan can also be very helpful. Sedums can act as an organic transition from lawns into the rest of the landscape or garden. Planted along walkways, they soften the hard edges of the material used. In beds and borders, they can benefit the plants around them by shading their roots and keeping the ground cooler. Because they are rarely aggressive, groundcovering sedums are unlikely to become a problem in a border. Owing to their stature, they are best near the front as edgings or weaving in and among lower-growing or clump-forming perennials so they don't get completely lost from view.

A closer look at two of the most popular sedums used as groundcovers illustrates some of these points. One is an old stand-by that's been around for years, *Phedimus spurius* 'Schorbusser Blut' (dragon's blood). Actually, today I prefer and always recommend its improved form 'Fuldaglut' (fireglow). As you might guess, fiery red leaf color is the main attraction with either plant. The other favorite is an excellent newer sedum with

Fall color on *Phedimus middendorffianus* can be some of the most intense red you can imagine.

A groundcovering sedum backed by coneflowers edges a stone path.

gold foliage, *Petrosedum rupestre* 'Angelina'. Leaf color in both is easy to contrast with surrounding plants. Both plants are super-adaptable, tolerating dry to average heavy or poor soil and prospering in everything from full sun to part shade. Their texture also contrasts easily with surrounding plants. Finally, both of these make outstanding groundcovers because their spreading habit not only softens sharp edges and fills in to exclude weeds, but also allows more plants to be made easily by division or rooted stems. If you wish, these extensions can also be separated from the original plant and planted in another spot.

Groundcovering sedums offer much more than just green. Many begin the growing season in shades of green or blue, but develop darker or red coloration as the summer progresses, and turn bright red or bronze or other gorgeous hues as winter approaches. Deciduous sedums don't necessarily start out colorful, but fall's cool weather brings out more vivid hues. A few have truly unbeatable fall color, so if you mass them as a groundcover, you get quite a show. *Phedimus ellacombeanus* turns orange-red, *P. middendorffianus*, vivid cherry red, and *P. sichotensis*, orange to scarlet.

Over the winter, some of the more half-hardy or tender-yet-evergreen sedums turn colors, too. Standouts among this group are *Sedum ×luteoviride*, which turns bronze; *S. moranense*, plum red; and *S. tetractinum* 'Coral Reef', coral red.

If you desire a certain foliage color in your groundcover sedum choices, consult the lists in this chapter. Expect that there are often seasonal variations or changes, and the setting—especially the exposure, more sunlight or less—can influence coloration. In my view, some variability adds to their value as groundcovers and only enhances the beauty of a garden throughout the year.

BLUE FOLIAGE

Sedums that have foliage in the blue range offer interesting groundcovering possibilities. Blue has a cooling effect that is restful to the eye and could be welcome in certain areas of landscaping, such as in a quiet corner where there's a seat, in a courtyard, or at an edge or boundary to make what designers call a "soft stop." A very popular blue groundcover is *Petrosedum rupestre* 'Blue Spruce', which has needlelike foliage reminiscent of a true evergreen shrub. Much of the time I see it planted around the base of evergreen shrubs, or in conjunction with low sprawlers such as juniper. Alternatively, you could combine a blue-hued sedum with similar ones of contrasting colors for added dimension; *Petrosedum rupestre* also comes in green and gold.

SILVER TO GRAY-GREEN FOLIAGE

Like other succulents, it seems that many sedums, both of the groundcovering or stonecrop kinds as well as the bigger hylotelephiums, have softer, muted foliage—leaves in the silver to gray realm. Sometimes this is due to a naturally occurring waxy or powdery coating over darker foliage. Other times fuzzy silver-leaved types have tiny hairs that protect against heat and drought. In either case, the coating gives silver its special reflective qualities—perhaps most obvious in early morning or late afternoon light, or when brightening dimmer corners.

These sedums are splendid combiners in practically any garden setting. They help green companions stand out better. They make hot hues look less garish and more stylish. Pastels look prettier.

PLUM, RED, AND PURPLE FOLIAGE

As for purple or plum foliage, a sedum in this color range can look sophisticated or sultry. It can act as a visual anchor, giving an area solidity and a backdrop against which flowers and foliage of all other colors may gain a lot of drama. Fortunately in the world of sedums, the palette is increasing. *Hylotelephium* 'Bertram Anderson' is a reliable form, but some of the new ones offer different textures, such as *H.* 'Plum Perfection' with its smaller foliage or *H.* 'Cherry Tart', with its thicker, rounded leaves.

A popular groundcover, *Petrosedum rupestre* 'Blue Spruce' features blue needlelike foliage and yellow flowers on vigorous spreading plants.

Groundcovers with Blue Foliage

Hylotelephium telephium subsp. *ruprechtii* 'Hab Gray'
Hylotelephium 'Thundercloud'
Petrosedum montanum subsp. *orientale*
Petrosedum rupestre 'Blue Spruce'
Petrosedum sediforme
Rhodiola pachyclados
Sedum clavatum
Sedum laxum
Sedum morganianum
Sedum oregonense
Sedum pachyphyllum
Sedum treleasei

In full sun, the glaucous blue-green, almost silver foliage of *Hylotelephium sieboldii* dons a red edge.

Groundcovers with Silver to Gray-Green Foliage

Hylotelephium 'Carl'
Hylotelephium ewersii subsp. *homo-phyllum* 'Rosenteppich'
Hylotelephium 'Pure Joy'
Hylotelephium sieboldii
Hylotelephium telephium subsp. *rupre-chtii* 'Hab Gray'
Hylotelephium 'Thundercloud'
Orostachys boehmeri
Sedum dasyphyllum
Sedum hispanicum var. *minus*
Sedum oaxacanum
Sedum palmeri

Groundcovers with Plum, Red, and Purple Foliage

Hylotelephium 'Bertram Anderson'
Hylotelephium cauticola 'Lidakense'
Hylotelephium 'Cherry Tart'
Hylotelephium 'Plum Perfection'
Hylotelephium telephium 'Black Beauty'
Hylotelephium telephium 'Cherry Truffle'
Hylotelephium telephium 'Purple Emperor'
Hylotelephium ussuriense 'Turkish Delight'
Phedimus spurius 'Fuldaglut' and other cultivars
Phedimus spurius 'Red Carpet'
Sedum album 'Coral Carpet'
Sedum ×*rubrotinctum* 'Aurora'
Sedum spathulifolium subsp. *purpureum*
Sedum stahlii

The plum purple foliage of *Hylotelephium telephium* 'Purple Emperor' with the gold-leaved *Petrosedum rupestre* 'Angelina' is striking.

YELLOW TO GOLD FOLIAGE

Gold-leaved sedums are gorgeous and worth considering where you want a fresh, warm look. One I often see used for groundcover regardless of the situation is *Petrosedum rupestre* 'Angelina'. It's durable, it spreads out well, and it's colorful. Yes, it spends much of the growing season earning its "gold" name and adding sparkle to the landscape. But, being evergreen, it has winter interest; cold weather causes its foliage to turn bronze with copper highlights. Some of my favorite combinations involve siting it with the red or plum-purple foliage of *Heuchera* 'Obsidian' or *H.* 'Plum Pudding.'

If you like this look but have less space to work with, consider *Sedum sexangulare* 'Golddigger'. While it is not as bright, this chartreuse form of the species turns toward gold in late spring to early summer, and then is joined by golden yellow flowers. Its overall texture is finer, with stems only around ¼ inch (6 mm) wide. 'Golddigger' forms spreading mats 2–3 inches (5–7.5 cm) tall and more than 12 inches (30 cm) across. Like many sedums, it will tolerate part shade. More shade seems to diminish the gold coloration.

ORANGE TO CORAL FOLIAGE

While there aren't a lot of sedums that naturally come in orange, coral, or similar warm hues, the ones that do can be spectacular. They're particularly striking when you mix them with one another, or toss in a totally different hue—blue, for instance—for contrast. Orange and lime green or chartreuse is also a wonderful pairing.

GREEN FOLIAGE

If you are seeking a sedum that has primarily green foliage for most of the growing season, several options are available (see list on page 23). A mature patch is a fine alternative to lawn grass and other, more traditional green-leaved groundcovers.

Sedum tetractinum 'Coral Reef'

Groundcovers with Orange to Coral Foliage

Sedum adolphii 'Coppertone'
Sedum emarginatum 'Eco-Mt. Emei'
Sedum japonicum var. *pumilum*
Sedum kimnachii
Sedum makinoi 'Limelight'
Sedum polytrichoides 'Chocolate Ball'
Sedum sexangulare 'Red Hill'
Sedum tetractinum 'Coral Reef'

Sedum sexangulare 'Golddigger' has fine-textured foliage that turns gold in summer.

Groundcovers with Yyellow to Gold Foliage

Hylotelephium erythrostictum 'Mediovariegatum'
Petrosedum rupestre 'Angelina'
Sedum acre 'Aureum'
Sedum japonicum 'Tokyo Sun'
Sedum makinoi 'Limelight'
Sedum makinoi 'Ogon'
Sedum mexicanum
Sedum mexicanum 'Lemon Ball'
Sedum sexangulare 'Golddigger'

The glossy green foliage of *Sedum* 'Sublime', introduced in 2008 by Intrinsic Perennial Gardens, makes a fine addition to the perennial bed or border.

Groundcovers with Green Foliage

Phedimus ellacombeanus cultivars
Phedimus hybridus 'Immergrünchen'
Phedimus kamtschaticus cultivars
Phedimus kamtschaticus var. *floriferus* 'Weihenstephaner Gold'
Phedimus spurius cultivars
Sedum acre cultivars
Sedum album var. *micranthum* 'Chloroticum'
Sedum divergens
Sedum mexicanum
Sedum oreganum
Sedum sexangulare cultivars
Sedum tetractinum

VARIEGATED SEDUMS

Sedums with leaves of mixed colors may serve various uses. In a potted display, they can be a focal point or offer contrast, or at least attract curiosity. In a mixed border, a sedum of a different color can "earn its keep" when it is not in flower, bringing fresh interest. Planted as edging or in larger groups or patches, variegated ones really grab attention.

Sedums, like so many other plants, can change or mutate based on their growing conditions. Sometimes an all-green species will develop a desirable variegation or color change. This has led to some interesting new plants from time to time.

Variegated foliage is defined as leaves that have at least one color contrasting with the natural or usual green. Among sedums, a desirable one would be a different-colored edge or center to the foliage. For example, *Hylotelephium* 'Herbstfreude' developed or "sported" to a yellow-bordered leaf. You can now buy such a plant under the name *Hylotelephium* 'Lajos'. An alert nurseryman or home gardener can propagate this different look to make a new-looking plant.

Hylotelephium 'Maestro', a sport of 'Matrona'.

Variegated plants have some drawbacks. Because they have less chlorophyll, they can be less resilient, sometimes also less winter-hardy. And in the case of sedums, "reversion" is a common issue. The all-green foliage tends to be faster-growing than the variegated parts, and the green returns and takes over. To keep these plants "true," watch carefully and if you see this starting to happen, act promptly. Cut out the all-green pieces as close to the base of the plants as possible, a process called rouging.

Occasionally variegation occurs as result of a plant virus. The way to tell is if the new color appears as a splash or irregular streaks. These sedums are not stable, nor are they healthy. The affliction could spread to other plants in your garden. (For more on viruses in sedums, please see page 202.)

A "sport" is a similar situation—also a genetic mutation—but refers to a shoot or new plant coming off a parent plant; the sport looks different in color or form. It can be separated off and grown on its own, and if it appears worthwhile, propagated vegetatively to build up stock. For example, *Hylotelephium* 'Matrona' generated a darker-colored sport that was eventually offered to the gardening world as 'Maestro'. The original plant has gray-green foliage, plum stems, and pink flowers. 'Maestro' has darker, practically purple leaves and deeper pink flowers. In the case of *Petrosedum rupestre* 'Angelina', the entire plant sported to gold foliage.

Occasionally sedums form fasciated shoots or crests, which result in flattened stems and a wavy ridge of foliage at the top of the stem. These are a novelty, but can be quite interesting if you are building a collection. I have a crested form of *Sedum album* 'Coral Carpet' that I like for its more controlled growth and habit.

Hylotelephium 'White Tooth Shark', a sport of the all-green *H.* 'Herbstfreude', has a thin white edge, but reverts back to the original form easily.

Groundcovers with Variegated Foliage

Hylotelephium 'Beka'
Hylotelephium 'Elsie's Gold'
Hylotelephium erythrostictum 'Frosty Morn'
Hylotelephium erythrostictum 'Mediovariegatum'
Hylotelephium 'Lajos'
Hylotelephium sieboldii f. *variegatum*
Hylotelephium spectabile 'Pink Chablis'
Phedimus ellacombeanus 'The Edge'
Phedimus kamtschaticus 'Variegatus'
Phedimus spurius 'Tricolor'
Sedum anglicum 'Suzie Q'
Sedum lineare 'Variegatum'
Sedum makinoi 'Variegatum'

Phedimus kamtschaticus 'Variegatus'.

A bed of the "cloud sedum" *Hylotelephium* 'Pure Joy'.

CLOUD SEDUMS

The possibility for groundcovering is not limited only to the lower-growing, stonecrop-type sedums. Some of the newer upright-growing or border sedums will also fill the bill. These recent arrivals have become so compact and heavy flowering that an entire plant looks like one continuous flower; I like to call these "cloud types." They make a unique, undulating groundcover when planted close together. To get this effect, simply space the plants 12–15 inches (30–38 cm) apart on center. Examples include white *Hylotelephium* 'Thundercloud', bubblegum pink *H.* 'Pure Joy', and magenta pink *H.* 'Birthday Party'.

On a grander scale, say, in a grassy meadow planting scheme, these newer introductions would be appropriate where a groundcover type would get lost. Their bunches of contrasting texture and color would stand out against the mellow green, fine texture, and eventually straw yellow of the dominant grasses from summer into winter. In particular, their colorful fall appearance—in foliage and flower—would be highlighted in such a setting.

Sedums in Containers

In recent years, interest in container gardening has exploded. Containers are ideal for anyone with limited space, time, or money. They can also fit anywhere—on a patio or deck, on a porch, going up a set of steps, even adorning a fire escape outside an apartment in the heart of a city. People with bigger gardens sometimes tuck potted plants into their displays; the plants can be moved if they aren't doing well or if a new idea occurs. Sedums are an easy choice for this trend, not only because of their diversity but also because almost all of them are easy-care.

As you probably are well aware, plants in pots dry out quickly. The soil mix holding their root ball may not be very large and water drains through. The exposed sides of the container heat up in the sun or dry out in the wind, which further depletes any moisture. Terra-cotta or clay pots actually wick moisture away from root systems. For these reasons, some potted plants require daily watering—especially in the heat of summer. For busy gardeners, caring for containers can become a chore, but not with sedums. Sedums tend to be drought-tolerant and flourish

Sedums are an integral part of some of the most imaginative container plantings, like this one at the Missouri Botanic Garden.

Drought-tolerant *Hylotele-phium cauticola* 'Lidakense' will look good in a container from spring to fall.

in containers, including terra-cotta and clay ones. In fact, most of them like dry growing conditions.

It's popular and practical to combine sedums with other drought-tolerant plants to make an all-succulent container. With creativity, these are sure to look marvelous. Plus, they're low-maintenance. So in addition to recommending specific sedums to plant in containers, I will also list succulents that complement them.

But first, here are a few basic principles to help you compose successful potted displays. First, scale is important; make sure that the size of your plant generally fits the size of its container. Due to their natural form, many sedums are used as fillers or as spillers—start with young plants and be patient while they get established. Individual sedums may be grown to be admired solo, "as specimens." When you arrange several different ones in a container together, or combine sedums with nonsedum companions such as other succulents, consider color and texture and allow some space for them to grow into one another. Finally, if you choose sedums that are not winter-hardy in your area, be prepared to move them, container and all, inside for the winter.

Attractive potted displays of sedums take into account nonplant elements as well, that is, the container itself and the gravel or grit with which the plants are mulched. Troughs of stone, hypertufa, concrete, and the like confer a natural, rocklike look. These are ideal for fine-textured sedums (for instance, small-leaved, bun-forming varieties like *Sedum hispanicum* cultivars), perhaps in concert with some alpine and tight mounding plants. Clay, ceramic, and plastic pots sometimes come in colors or with decorations that may flatter or contrast well with the sedum display within—just be careful not to let the container itself distract from the beauty of the plants within. While most sedums (and

Many sedum plants will hang down from their container, like *Sedum mexicanum* 'Lemon Ball'.

Sedum morganianum used as a hanging plant in a mixed container.

A trio of sedums finds extra drainage in a strawberry pot: *Hylotelephium* 'Ruby Glow' (top), a red form of *Phedimus spurius* (center), and a green *Orostachys* (bottom).

Trailing Sedums for Containers

Sedum adolphii
Sedum clavatum
Sedum kimnachii
Sedum lineare
    'Variegatum'
Sedum mexicanum
    cultivars
Sedum moranense
Sedum
    morganianum
Sedum palmeri
Sedum stahlii
Sedum tetractinum
    cultivars
Sedum treleasei

succulents) like a fast-draining soil mix, you can always top off the project with a thin mulch of fine gravel, which comes in different sizes and even in different hues. It's a finishing, contrasting touch that helps retain a touch of soil moisture, discourages weeds, and can enhance your displays very nicely.

Offbeat containers are another possibility. Because sedums are easy-going, many will tolerate life in quirky containers. How about an old shoe or boot? A vintage tea kettle? A colander? A discarded paint or coffee can? If you are hesitant to plant directly in the container, simply tuck in a small pot. Whatever you use, remember to use a quick-draining or gritty soil mix. Also determine if it's possible for excess water to drain away from the roots—if not, create some drainage holes. If you have a wooded garden, consider planting right on top of or inside an old stump or log. I would recommend trying *Sedum ternatum*.

SEDUMS THAT HANG OR TRAIL

The best candidates for hanging baskets, windowboxes, or elevated pots or urns tend to be the more robust growers, ones with a naturally cascading habit. Burro's tail (*Sedum morganianum*) is the most common, but with age, *S.* ×*rubrotinctum* and *S. stahlii* trail, too.

A bold ceramic pot creates a colorful background for an array of sedums that includes *Hylotelephium* 'Sunset Cloud', *Phedimus takesimensis*, and *Hylotelephium sieboldii* f. *variegatum* growing out of the top, *Sedum oreganum* (front center), *Petrosedum forsterianum* (left background), and *Sedum ochroleucum* 'Red Wiggle' (right background). A sempervivum (bottom left) completes the composition.

Smaller trailing varieties like *S. diffusum*, *S. moranense*, and *S. oaxacanum* all work nice in small to medium-size pots.

SPECIMEN DISPLAYS

Some sedums are too small to gain attention out in the ground in the garden, but are sufficiently big enough to make a splashy show in a pot. These are worth seeking out and trying if you enjoy interesting and unusual potted plants.

Tree Sedums. Some of the Mexican species make long-lived specimen plants that grow up on woodylike stems. *Sedum dendroideum* and its relative *S. praealtum* look almost like a jade plant (*Crassula*), while *S. multiceps* at 3–6 inches (7.5–15 cm) tall is reminiscent of a miniature Joshua tree (*Yucca brevifolia*).

Stout Rosettes. Some that form thick-leaved rosettes, such as *Sedum lucidum* and *S. treleasei*, may remind you of their *Echeveria* relatives (and could be combined with them). They look great in small clay pots.

Bushy Choices. Certain other sedums, notably *Sedum furfuraceum* and *S. hernandezii*, mature into plants that look decidedly shrublike. They can make nice houseplants, too.

COLOR IN CONTAINERS

One of the most rewarding aspects of growing sedums in pots is the opportunity to create attractive combinations. So many of the smaller ones suitable for container life are colorful in both foliage and flower—there's a wide palette to choose from. You may already have favorite hues, green with red, for instance, or red with yellow, that you've enjoyed with other plants in other areas of your landscape. Apply the same successful or enjoyable partnerships here. A word to the wise, however: less is often more. Because so many of these plants are small in all aspects, you risk overwhelming or disguising their charms if you compose an overly busy or too-full display.

If you want, you can leverage the container itself to help show off your little sedums. A pot of contrasting color, such as bright blue, might turn out to be just the backdrop they need to stand out.

I also encourage you to move boldly beyond plain green or gray-green choices. Here's one good idea: mix blue foliage, such as *Sedum clavatum* or *S. pachyphyllum*, with orange foliage, such as *S. adolphii* 'Coppertone'.

CONTAINER COMPANIONS

As mentioned previously, a potted display that adds other plants to a sedum or sedums is successful if the growing requirements are a match. Succulents all like ample sunshine and quick-draining soil, so they are natural partners. Other drought-tolerant plants will work, too.

Remember that container displays are flexible. If, after a while, you notice a companion plant in a mixed pot is not doing well, or that the sedum is getting lost from view because its partner is getting too large or is a more aggressive grower, it's not a big deal to make changes. Either prune back rampant growth or swap out a plant altogether, replacing it with something that hopefully is more suitable. As you experiment and learn, you will develop combinations that work and delight you and garden visitors with their beauty.

Good container companions for sedums include *Armeria, Dianthus, Jovibarba, Lewisia, Rosularia, Saxifraga, Sempervivum,* and *Talinum.* Among the recommended tender companions are *Echeveria, Euphorbia, Graptopetalum, Pachyphytum, Portulaca,* and *Senecio.*

Sedums in Rock Gardens, Stone Walls, and Crevices

Because their native habitat is so often an exposed site of rocky soil, stonecrop-type sedums are naturals for rock gardens. Innovative gardeners and landscape designers have also experimented with tucking them into stone walls—more or less vertical surfaces—and into crevices. It could be that their work will inspire you to try something similar somewhere in your own yard. There is a lot of great information available from the North American Rock Garden Society when it comes to building different types of rock gardens (see page 215 for the address and contact information); I'll just provide an overview here.

ROCK AND GRAVEL GARDENS

The bulk of the work in a traditional rock garden is in the setting-up. Once it is in place and the sedums and companion plants become established, it becomes low-maintenance. The best sites are in full sun, constructed on a slope that allows for good drainage. Most will consist of small mounding plants from mountainous regions, typically mulched with gravel. The goal is to be easy-care, in particular to greatly reduce or eliminate the need for weeding.

One variation is the so-called gravel garden. This is not a new idea, but gravel gardens are up-and-coming in the United States these days. They tend to be built on flat ground. The entire area has the plants set on top of the prepared soil bed. Then the gravel is added like a mulch, up to and around the crown and soil ball of the plants to a minimum depth

Sedums are most at home in gravel gardens and other dry situations.

Gravel Matters

A TRIP TO a gravel yard or well-stocked garden center will show you many different options in terms of stone size, color, and even shape. In most cases, the role these small-size rocks play is as top-dress, or overlaid mulch; you want to actually grow your sedums and companions in soil, albeit quick-draining soil in most cases (a sandy mix suffices for many).

Smaller often looks the most natural. I often use a local pea or birds-eye gravel, which is rounded, measuring only around ¼ inch (6 mm) or less. Use bigger if your plants are larger.

A neutral color is a safe bet. For instance, I favor light gray gravel, a naturally mined angular stone from Wisconsin that resembles quartz. But some rock gardeners have fun with darker or lighter stones, red- or blue-hued ones, or mixes. While it's up to you, just bear in mind that you don't want an overly complicated display with too much going on. My advice is to keep it simple.

of 4 inches (10 cm). Some larger stones or boulders can be placed here and there to break up the flat surface of the bed.

In any event, once the design of a bed shape has been laid out, some sort of edging should be installed. This keeps gravel from spilling out into a path, lawn, or organically mulched areas. Use metal, stones, or pavers.

The plant palette is quite wide. You can, of course, devote an entire area to various sedums—I have a bed that houses my hardy sedum collection and it is beautiful. If you want to add companions, just remember to choose plants that are relatively drought-tolerant.

You can be traditional and add alpines. Be careful, though. Some of these are native to areas above the timberline and can be difficult to keep happy in a home garden; they can't tolerate high summer heat and must remain dry over the winter months. Sometimes you can get around these issues by raising fussy customers in pots. Then, either set them throughout the rock garden or submerge their pots into the ground for the growing season. Research the plants you admire, and decide from there which ones you are willing and able to grow.

For a different, meadowlike look, include low-growing native American grasses in your rock or gravel garden. These provide a handsome backdrop (often in all seasons) as

Sedums for Terraces and Walkways

Phedimus kamtschaticus
Phedimus spurius 'Dr. John Creech'
Rhodiola pachyclados
Sedum acre 'Aurea'
Sedum album var. *micranthum*
    'Chloroticum'
Sedum dasyphyllum
Sedum hispanicum
Sedum japonicum var. *pumilum*
Sedum lydium

With its tight habit that deters even weeds from breaking through, *Phedimus spurius* 'Dr. John Creech' makes a perfect edging along a path.

well as contrasting texture and form. In mine, I've had good luck with these shorter grasses: *Bouteloua curtipendula*, *B. gracilis*, *Koeleria cristata*, *Panicum leibergii*, *Schizachyrium scoparium*, and *Sporobolus heterolepsis* 'Tara'.

Other low-growing dryland plants are also welcome, with or without the grasses. I like to use native wildflowers, which are naturally tough and adaptable. They also have a nice informal look that works really well with both border and groundcovering (stonecrop) sedums like *Phedimus spurius* cultivars. Some companions I recommend are *Allium schoenoprasum* 'Rising Star', *Coreopsis lanceolata*, *Echinacea tennesseensis*, *Geum triflorum*, *Liatris cylindracea* and *L. punctata*, and *Oenothera macrocarpa*.

SEDUMS FOR TERRACES AND WALKWAYS
Thanks to their ability to survive in exposed, sunny locations with minimal water, stonecrop-type sedums can be a natural choice for planting among paving stones (flagstones, rocks, even bricks) in a terrace or walkway situation. They need to be able to withstand foot traffic, but not all can. Certain ones do fit this bill, though, and the species among them may even please you by self-sowing into adjacent cracks and openings

Sedums soften the sharp angles of cement steps and introduce color in this exposed, hot site.

A paver wall planted with a variety of sedums.

An example of a stone wall that has built-in pockets of soil to plant into, perfect for sedums.

Phedimus ellacombeanus naturally reseeds itself into cracks on a stone paver wall.

in the vicinity, over time giving the area a softer, more natural or established look. Help them out from time to time with a light watering and by yanking out competing weeds.

When shopping for plants at your local garden center, look for sedums with a Stepables program tag. These plants are able to tolerate various degrees of foot traffic.

STONE, ROCK, OR PAVER WALLS

Stone walls are part rock garden and part living or "green" wall. They can act as a rock garden with plants placed right on top or tucked in, that is, planted or seeded into holes in the face.

There are two main kinds. One is a dry stone wall that is not mortared and has limitations for height of only around 3 feet (90 cm) tall. The other is a taller wall made of either natural stone or pavers held together with mortar, and featuring some pockets for plants to be inserted. These pockets are best planned for and built into the wall during construction.

Such walls are best built with a natural slope to them. This serves two purposes. It provides stability, and it allows rainwater to penetrate the side where the plants are. They can be lodged against or supported by a natural embankment or hillside, or be free-standing.

While you will do the initial installation of plants in a stone-wall garden, you don't want this to become a constant maintenance chore. For best results, then, choose sedums that reseed easily on their own. Over time, they should fill in the display, including amazing you by working their way into even the smallest cracks. Here are some excellent candidates: *Phedimus ellacombeanus*, *P. kamtschaticus*, *P. spurius*, *Sedum acre*, *S. album*, and *S. sexangulare*. To add interest to a sedum rock wall planting, include some alpines that will also reseed into crevices, such as *Aurinia saxatilis*, *Aubrietia deltoidea*, and *Arabis caucasica*.

A design issue you may confront as you aim to "landscape" your stone wall is softening the hard surfaces. A remedy is to plant trailing sedums on top so they can hang over. Two that work great are dusky rose-plum *Hylotelephium pluricaule* var. *ezawe* and gray-silver *H. ewersii* subsp. *homophyllum* 'Rosenteppich' (rose carpet). Side by side, these two have very similar form, size, and leaf size, but their differing colors against the repeating surface of the stones create a gentler picture. This same effect could also be achieved by combining the equally fine-textured and spreading mat forms of silver-leaved *Sedum hispanicum* var. *minus* with plum-colored *S. hispanicum* var. *hispanicum*. Another choice that is comparable in size and texture but is green with red stems and highlights is *S. lydium*.

CREVICE GARDENS

Crevice gardens are a form of rock gardening that is gaining popularity. These unique displays mimic sheer flat-rock surfaces, with slits of rocky soil to plant into. They are created by placing flat rock vertically, with slices of area between each slab for plants. There are nice examples at Allen Centennial Gardens on the University of Wisconsin at Madison campus and at Denver Botanic Gardens.

A crevice garden in Allen Centennial Gardens on the University of Wisconsin-Madison campus.

As for suitable plants, keep them in scale, that is, small. In my own crevice garden made of flagstones, I have tucked in petite *Sedum dasyphyllum* and *S. japonicum* var. *pumilum*. For a companion, I added a short flowering pussytoes (*Antennaria neglecta* 'Pewee'). The pussytoes and *S. dasyphyllum* make a pleasing combination with the gray stone, while *S. japonicum* var. *pumilum* picks up some of the rust orange tones that are speckled on some of the neighboring stones. Small forms of *S. album* would also be appropriate. *Sedum album* var. *micranthum* 'Chloroticum' has bright green foliage all year. *Sedum album* 'Fårö' is the smallest cultivar and quite appealing; it turns red and green in fall, and becomes entirely red in winter.

These coconut coir trays are one example of a greenroof tray system.

Sedums for Greenroofs

Today, greenroofs are popular, or certainly cutting-edge as a concept for sustainable construction as well as gardening. Sedums are ideally suited for them and, indeed, a staple. Interestingly, the Chinese have been growing *Orostachys chanetii* (synonym *Sedum chanetii*) on rooftops for hundreds of years, maybe even centuries.

The practical benefits of greenroofs are many. Most such roofs are not just ornamental; they've been installed for utilitarian reasons. In large cities where runoff into already overburdened storm and sewer systems is a major concern, a greenroof can retain 60–100 percent of the rainwater. It also helps cool a building on hot days. And it may extend the life of a roof by more than 50 percent due to less ultraviolet exposure and less temperature fluctuations—thus also reducing costs over time. Finally, a greenroof provides habitat for native wildlife, including birds, bees, and other beneficial insects. It can create pockets of biodiversity and connections to natural and ground-level habitats for various creatures.

Most greenroofs are built atop new construction to allow for proper reinforcement to carry the extra weight. Flat roofs are most common since greenroofs are mostly still built on commercial buildings. For residential applications—in case you are thinking of trying this at home—consider a design that incorporates the greenroof into an outdoor living space like a deck or patio. Alternatively, one could be installed on a garage, shed, or playhouse roof. Bear in mind you're going to need access!

To meet the plants' needs, greenroofs are constructed in certain ways. There are a number of excellent books out on the topic, and/or you can hire a firm experienced with

Greenroofs can be planted using plugs.

Greenroofs can be planted using stem cuttings.

This 3-acre (1-hectare) "sedum sod" greenroof on a FedEx building at Chicago's O'Hare International airport was installed in just a few weeks by Intrinsic Land-scaping, my brother Kurt's company.

The greenroof at Chicago Botanic Garden in summer.

The greenroof at Chicago Botanic Garden in spring. The roof has multiple soil depths of 4, 6, and 8 inches (10, 15, and 20 cm), with sedums generally planted in 4 inches (10 cm) of soil.

Eight months after sedum cuttings were installed on this roof at Peggy Notebaert Nature Museum in Chicago, Illinois, approximately 80 percent of the roof is covered.

greenroof installation in your climate. Briefly, though, a shallow mix of 3–5 inches (7.5–13 cm) of soil suits most sedums best; deeper installations allow deeper-rooted plants and thus a broader plant palette. Also greenroof soils are specialized to minimize shrinkage and discourage them from being blown away. In the Pacific Northwest, they use pumice, while in the Midwest and much of the rest of the country, baked clay or expanded shale is common. The other media components include sand, peat moss, perlite, compost, and additional organic planting materials.

There are a few installation methods. In the most traditional, long used in Germany and other European countries, the main components are constructed right on top of the roof. If there is a specific design, it is planted with plugs; otherwise, the roof may be hydro-seeded with sedum cuttings. Another way is to install pre-grown trays. The trays are usually plastic but sometimes include coconut coir. A pre-grown vegetative mat, also known as "sedum sod," is a third possibility.

Sedums for an Extensive Greenroof

These plants need 3–5 inches (7.5–13 cm) of soil in Zone 5.

Hylotelephium cauticola 'Lidakense'
Petrosedum rupestre cultivars
Phedimus hybridus 'Immergrünchen'
Phedimus kamtschaticus cultivars
Phedimus kamtschaticus var. *floriferus* 'Weihenstephaner Gold'
Phedimus spurius cultivars
Phedimus takesimensis 'Golden Carpet'
Rhodiola pachyclados
Sedum acre cultivars
Sedum album cultivars
Sedum pulchellum
Sedum sexangulare cultivars

Most greenroofs take one to three years to become established. Expect to do some maintenance. In the beginning, you should fertilize regularly, given the typically lean, low-fertility soil used. Just as with any new garden, keeping after the weeds until the desired plants fill in is important. Last but not least, since you're dealing with a roof, always keep drainage channels clear.

Sedums for Carpet Beds and Living Walls

One of the most unusual and exciting ways sedums can be used is in carpet beds on the ground or in one-sided vertical beds that form or are secured to a wall or strong fence. Either way, the project is started by creating a frame to contain the plants; the frame may be wood, brick, or other sturdy edging material. This is filled with soil and then an array of plants, often but not always in a strict geometrical arrangement. The result looks like a framed, living picture.

Some plants that lend themselves to carpet-bed use include *Sedum spathulifolium* subsp. *pruinosum* 'Cape Blanco' and *S. dasyphyllum* for their strongly silver foliage, *S. album* 'Coral Carpet' and *Phedimus spurius* 'Fuldaglut' (fireglow) for their red foliage, and *Petrosedum rupestre* 'Angelina' and *S. japonicum* 'Tokyo Sun' for their golden yellow foliage.

Living walls take this concept and display it vertically, sometimes on a larger scale. Low-growing, drought-tolerant sedums are good choices for these because they not only

A formal carpetlike planting of *Sedum mexicanum* 'Lemon Ball' in the Missouri Botanic Garden's Victorian Garden.

An all-sedum living wall in the Grunsfeld Children's Growing Garden at Chicago Botanic Garden.

Sedums are ideal plants for a framed vertical garden.

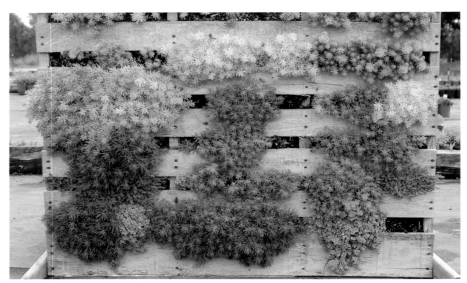

A wooden pallet planted with sedums becomes a sturdy privacy panel.

A living picture can be created by using a framed box of soil and planting it with sedums and other succulents.

Sedums combined with creeping thymes on a living wall at Chicago Botanic Garden.

stay in bounds, but also don't demand intensive maintenance. That said, supplying what they need to stay alive and look good could be tricky. You want a well-drained soil, yet there must be constant moisture available to both plants and soil (traditional peat moss-based mixes hold too much water and have issues with shrinkage). Additionally, to keep plants healthy and prevent diseases, the substrate ought to dry down between waterings. Ideally the living wall panels are planted and established on the ground and then hung once the plants fill in.

Unfortunately hydroponics, where the plants are maintained by water, not soil, doesn't work well for sedums and other succulents in these constructions. (Hydroponics is used for living walls of tropical plants—sometimes seen inside office buildings or shopping malls.) Two other options seem to work, however. Absorbent felt "vegetation mats" (an idea pioneered by French designer Patrick Blanc) suffice. A large plastic or metal tray lining the frame of the display, or rather nesting right inside its bounds, also works. The tray must have drainage holes, though, or root rot is a threat. Either of these methods still requires attention to irrigation.

The living wall plant palette depends on climate. In an extreme climate with very cold, wet winter weather, the number of plants that give winter interest is limited. These include evergreen sedums such as *Phedimus kamtschaticus* var. *floriferus* 'Weihenstephaner Gold', *P. takesimensis* 'Golden Carpet', *Petrosedum rupestre* cultivars, and semievergreen forms like *Hylotelephium anacampseros* and *Phedimus spurius* cultivars that hang down on stems tipped with rosettes. In warmer climates, of course, the palette expands considerably.

Sedums for Very Dry Conditions

While many sedums tolerate dry situations, certain species can withstand excessively dry, hot (xeric) conditions like those found in the Southwest, Far West, or their native Mexico. They can go for weeks without any water. These are also appropriate if you are a neglectful waterer or like to display sedums in pots on a hot patio or deck where less-tough ones will simply bake in the sun.

There is a good reason these particular sedums as well as certain other succulents can survive such conditions. It's a phenomenon known as "CAM," or "Crassulacean acid metabolism," which allows them to conserve water in their leaves. Most plants open their stomates (porelike openings on the leaf surface) during the day. "CAM" plants, however, only open their reduced stomates at night; the stomates remain closed during the heat of the day, which reduces water loss from transpiration.

Thick, succulent foliage is a clue that a plant is going to be tough. For instance, the jellybean stonecrops are champions in heat and drought. This is a small but important group that includes *Sedum furfuraceum*, *S. hernandezii*, and *S. stahlii*, along with their hybrids. Species like *S. adolphii*, *S. clavatum*, *S. lucidum*, and *S. treleasei*, with leaves at least ¼ inch (6 mm) thick, laugh at the heat.

Other desert natives are exceedingly durable. The tree sedums (*Sedum dendroideum* and *S. praealtum*) along with their shrubby counterparts (*S. confusum* and *S. kimnachii*) make long-lived, trouble-free container plants. If you live in an area with little freezing, you can grow them in the ground and they will survive year-round.

What about border sedums? Are any of them extra-drought-tolerant? As a rule, they are not, but you might be surprised to learn that there are a couple of exceptions. I have never seen *Hylotelephium cauticola*, *H. populifolium*, or *H.* 'Plum Perfection' wilt, but I don't grow them in the desert either.

Hot, dry conditions can also actually improve a sedum's appearance; some develop summer highlights, that is, experience a color change. Examples include the thick-fingered foliage of *Sedum morganianum*, *S.* ×*rubrotinctum* and its cultivar 'Aurora', and *S. pachyphyllum*. The tiny, shiny leaves of *S.* ×*rubrotinctum* turn super bright red with hints of orange and green, while 'Aurora' has the added characteristic of creamy white new growth and a brighter cherry red color. *Sedum pachyphyllum* gets orange to red tips against the blue leaves—they look like blinking lights on the ends. Cleverly named *S. adolphii* 'Coppertone' also improves with heat, turning on the orange lights for summer.

Excessive heat and dryness do not restrict the potential for attractive displays. Try *Sedum palmeri*, which looks just like a miniature palm tree on up to 1-foot (30-cm) cane-like stems, underplanted with *S. diffusum* 'Potosinum' (which appreciates a little shade) or *S. moranense* (which turns purple-red over the winter). For a hot-colored show in a hot spot, try *S.* ×*rubrotinctum*, which makes a hot combo of red and green all by itself. Or you can cool it down by adding some blue fine foliage like *Senecio mandraliscae* or *Euphorbia myrsinites*.

Sedums for Very Dry Conditions

Sedum adolphii cultivars
Sedum allantoides cultivars
Sedum clavatum
Sedum dendroideum
Sedum lucidum cultivars
Sedum oaxacanum
Sedum pachyphyllum
Sedum ×*rubrotinctum* cultivars
Sedum suaveolens
Sedum treleasei

Sedum ×*rubrotinctum* 'Aurora' foliage turns cherry red in hot and dry growing conditions.

Sedums for Shade

Although sedums are well-known sun-lovers, there are some exceptions, ones that can be grown in shade. In fact, shade can be the determining factor in keeping some of these types alive. In regions with hot, humid summers, the following species need to be protected on the hottest days with some shading: *Phedimus forsterianum*, *Sedum makinoi* cultivars, *S. polytrichoides*, *S. glaucophyllum*, and *S. japonicum* 'Tokyo Sun'.

The sedums that like shade seem to also prefer some moisture to be happiest. *Sedum sarmentosum* is a good example. Even though it is from China, it is often used in Japanese gardens. Another is an eastern United States native that hails from the Appalachians, *S. ternatum*. It has rich green, rounded foliage with white sprays of flowers in spring. A plant from the Rocky Mountains, *Rhodiola integrifolia* will surprisingly tolerate damp soil if summers are cooler like its native haunts.

The damp Pacific Northwest brings us some interesting native selections as well. Two are deep green-leaved species that are related, *Sedum divergens* and *S. oreganum* (over winter, both tend to take on plum red highlights). Another great choice is *S. spathulifolium* cultivars. These want a rocky or fast-draining soil and a cool location. Many of the subspecies and varieties have chalky silver or red foliage that would play off the deep greens of *S. divergens* and *S. oreganum*.

A sedum with hostas? Believe it or not, it can be done, and the extreme contrast between the leaf sizes and forms is very striking. I saw this once in a shade garden. *Sedum hispanicum* var. *minus* had been used as an edging plant. Not only had it adapted, it had also seeded and spread around vegetatively.

In hot, humid coastal areas, *Sedum mexicanum* and its gold cultivar 'Lemon Ball' will fill the bill. I've even seen them sharing this job with the popular groundcovers *Liriope spicata* and *L. muscari*, the sedum's needlelike foliage provided intriguing contrast to the liriope's evergreen blades. Create a similar green-and-gold effect with *S. makinoi* 'Ogon' and the green species, or the lighter green form *S. makinoi* 'Limelight'. Or you might consider *Phedimus stoloniferus* for its brighter green, rounded foliage set next to red stems on nearly ground-hugging plants. In cooler climates, use *S. acre* and *S. acre* 'Aureum'.

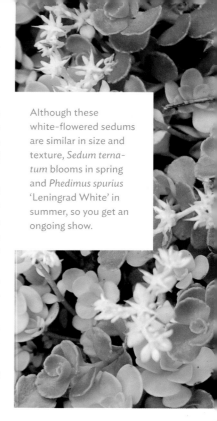

Although these white-flowered sedums are similar in size and texture, *Sedum ternatum* blooms in spring and *Phedimus spurius* 'Leningrad White' in summer, so you get an ongoing show.

The adaptable *Sedum sexangulare* and its gold form 'Golddigger' blend easily yet provide contrast.

Sedum sarmentosum is a rapid spreader in moist shade. Planted around boulders and evergreens, it will cradle neighboring plants.

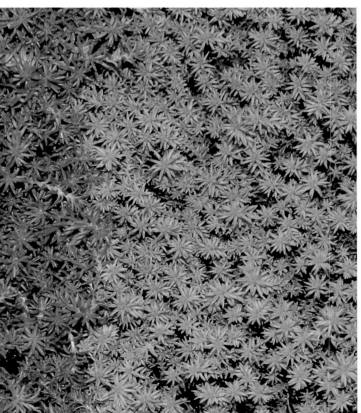

For strong contrast in partial shade settings, interplant plain green *Sedum mexicanum* with its gold cultivar 'Lemon Ball'.

One of the smallest sedums,
Sedum album 'Fårö'.

Sedums for Terrariums, Fairy Gardens, and Bonsai Displays

Some of the smallest-leaved sedums take time and care to fill in, but in the right spot can have a dramatic effect. These miniatures are ideal for terrariums, which, believe it or not, are making a comeback. They are also a charming choice for "fairy gardens," which are increasingly popular. Bonsai enthusiasts like them for filling in around the bases of their tiny trees. If any of these concepts entice you, by all means look up information on the Internet or acquire some of the new books on these topics—you'll find plenty of creative ideas and good plant suggestions.

The most important factor in these specialized settings will be appropriate soil. In particular, terrarium containers can become damp and humid inside, which is not ideal for many sedums. So don't use the usual soil mixes; use a fine rocky blend instead.

Varying color and texture makes a display more interesting, but try not to be too ambitious or the result will look "too busy."

A newly planted sedum wreath.

These planted globes are a nifty accent for the formal planting they are set in, the Victorian Garden at Missouri Botanic Garden. Here, *Sedum japonicum* 'Tokyo Sun' is combined with red hens and chicks, *Sempervivum*.

Sedums for Living Wreaths, Topiary, and Arrangements

If you use an appropriate base, it's easy and delightful to create a living wreath or a topiary using sedums and related plants. Take a wire frame in the shape you desire and stuff it with pre-dampened sphagnum moss or other water retentive yet airy soil mix. Some hobbyists or florists add another step: they wrap the entire thing loosely in thin plastic nonadhesive tape, which holds the soil mix in place. Then they make small holes in it at intervals and poke in the plants. As a finishing touch, some thin green florist's wire can truss the entire construction in place. To water a living wreath or a topiary of sedums, simply soak it in a pan or sink of water, then let the excess moisture drain off.

The sedums most suitable for wreaths are the thick-leaved, rosette-forming ones, such as *Sedum adolphii*, *S. clavatum*, *S. lucidum*, *S. pachyphyllum*, *S. ×rubrotinctum*, and *S. treleasei*. You can also tuck in other succulents like echeverias and hens and chicks. For globes and other topiary shapes, try *S. album*, *S. dasyphyllum*, *S. grisebachii*, *S. japonicum* 'Tokyo Sun', *S. lydium*, *S. spathulifolium*, and red hens and chicks.

UNDERSTANDING
SEDUMS

Petrosedum ochroleucum 'Red Wiggle' has fleshy leaves that hug the stem.

S

Sedums comprise a large and wonderfully varied group of plants. Some have been in cultivation a long time, while others are more recent introductions, either discovered in nature or culled by a discerning horticulturist in a garden or nursery setting. Although the botany of these plants appears somewhat complicated at first glance, I hope to guide you successfully through what turns out to be a sensible and helpful classification system.

What Is a Sedum?

While I like to call "sedum" a family, to speak correctly the plants belong to the botanical family Crassulaceae. You may be familiar with the flagship member of this family, jade plant (*Crassula ovata*) or with colorful-flowered kalanchoes as houseplants or as mild-climate landscaping favorites; neither are sedums, but they have the characteristic succulent leaves—that is, fleshy, moisture-conserving—that all members of this family share.

It is their foliage we often notice the most, but sedums are distinguished within this larger family by their flower form. Here, we observe that individual flowers have an equal number of parts, so if there are five petals, the sepals and male and female parts will also be in multiples of five. Plants in the actual genus *Sedum* usually have twice the number of stamens (male flower parts). A number of closely related plants, some of which are also called "sedums" by gardeners and nurseries but go under other genus names, are also

Sedum emarginatum features succulent, moisture-conserving
leaves, which all sedums share. Note the notched leaf pattern, an
identifying characteristic in this particular species.

grouped with them by virtue of being in the Crassulaceae and having succulent foliage and equal-parted flowers.

Since 2000, taxonomists have been busy making distinctions within the plants placed in this group. Some of this careful study has been aided by the advent of sophisticated DNA research (which, you may know, has also caused shuffling and name changes within other plant groups). All told, there are currently 33 genera in this family. Taxonomists then go on to place 400, and possibly up to 800, species here. In this book, I have winnowed down a selection of what I consider to be the 150 most useful sedums now available, including a number of exciting newer ones.

The Genera

Not all sedums belong to the genus *Sedum* anymore. Plants in five subgenera that used to be considered part of the genus have characteristics now judged to be sufficient to separate them from *Sedum*. These newer genera are *Hylotelephium*, *Petrosedum*, *Phedimus*, *Rhodiola*, and *Sinocrassula*. In this book, I shall focus mainly on the first four. There are also a couple of additional genera that taxonomists consider very closely related; a representative few of these are also included. Here is a general overview to help orient you.

Sedum is derived from the Latin *sedere* (to sit), although it was thought to have possibly come from *sedare* (to calm). It was chosen by Carl Linnaeus in 1737 to represent the genus because the name was commonly used for centuries already. The Greek word *aizoon* also referenced these plants. It means "ever-living" or "always, eternal" and reflects the natural toughness and longevity of many sedums.

For the most part, *Hylotelephium* species are easily distinguished from other sedums by their taller upright and leafy stems, broad, slim leaves, compact, dome-shaped flowerheads, and stems that die back to a distinctive tuberous rootstock. While there are a fair number of selections with creamy white petals and flower parts, none are yellow. Also, *Hylotelephium* carpels (female flower parts) are always upright. Gardeners will recognize the sedums that

Hylotelephium telephium 'Xenox' is a classic border sedum, with broad, thin leaves and compact, domed flowerheads.

Hylotelephium sieboldii has the broad, thin leaves typical of this genus, but it also exhibits a three-part (ternate) leaf pattern, which is unique to the species and a characteristic it passes on to its cultivars.

Aptly named *Petrosedum rupestre* 'Green Spruce' also has needlelike leaves.

Petrosedum rupestre 'Angelina' has needle-shaped foliage.

Phedimus takesimensis 'Golden Carpet' with the top half of the leaves notched. Notice also that the leaf margins are very pronounced.

New shoots emerge from the top of the aboveground rootstock of *Rhodiola rosea*.

Sinocrassula indica var. *yunnanense* in its rosette stage.

have been transferred into this genus as the classic "border sedums," which includes the ever-popular 'Herbstfreunde'.

Petrosedum species are readily recognized for their needle-shaped foliage. Their flowers are yellow to cream-colored. In the past, some of these were considered *Sedum*—in the so-called "rupestre" group. As taxonomic research has advanced, *Petrosedum* has emerged as a clear genus with ten species and three subspecies. Although this may appear to be a surprisingly small number, it does include some of the most popular cultivated sedums, in particular *P. rupestre* 'Angelina'.

Phedimus species have thinner, flatter leaves in contrast to true *Sedum* species, which have finger-shaped or thickened leaves. Also, the leaf margins are distinctly toothed, or serrated (that is, unevenly toothed like a saw), as opposed to the smoother or "entire" margins of *Sedum* species. *Phedimus* species are primarily herbaceous perennials, although some do get woody at their bases. In winter, the plants tend to lose most of their foliage, leaving bare stems and sometimes retaining ragged-looking rosettes of foliage at their tips. They revive and generate attractive new growth the following spring.

Rhodiola is a genus of many species distributed primarily in the mountains of the United States and China. Though at a quick glance the plants might look like sedums, they differ in three main ways. The leaves are thinner and the stems grow out of the top of their thick rootstock that is partially aboveground rather than coming from underground. (The common name is roseroot, which comes from the sweet smell of the rootstock.) As a result, the plants appear multistemmed, like a small thicket. Only a handful of rhodiolas are commonly grown.

Sinocrassula species are tender plants from South China and northern Myanmar (*Sinocrassula* means "Chinese crassula"). They are relatively small, rosette-forming plants with dark foliage reminiscent of a sedum. Their distinctive little flowers have a single row of stamens compared to the two rows typically found on *Sedum*; these are carried in panicles and are white or deep pink to red. I have only included one good *Sinocrassula* in this book, since these plants are not generally available in North America or Europe. Very occasionally, however, others are available in the trade, sometimes mislabeled and sold as sedums.

Last but not least, to add depth to the plant palette here and to acknowledge a couple more plants previously classified as sedums, I have also included an *Orostachys* species and a *Prometheum* species. The *Orostachys* name has been commonly accepted into general use. While the rosette foliage of *Orostachys* is similar to that seen on some sedums, the unique spike-shaped flowerheads set them apart. Yet their individual flowers can resemble those of *Hylotelephium*, so they do belong here. The genus *Prometheum* has only two species (both still often sold as *Sedum*). Both of these are biennials that form small rosettes of fuzzy foliage. They are monocarpic, meaning that they die after flowering.

Where Sedums Grow in the Wild

With so many different species, it is curious to me that the majority of sedums come from the Northern Hemisphere. As it turns out, there are some South American species, but they are just beginning to be named and classified. Time will tell if any are worthy of our gardens.

In fact, there are three main areas of heavy distribution. The Mediterranean region has around 100 species, trickling into northeast Africa, the Middle East, and the mountains of Europe and Scandinavia. The Himalayan Mountains and surrounding areas into the Pacific Rim are home to over 200 species. Quite a number of these turn out to be coastal plants or native to various Pacific islands. Lastly, another 100 species hail from Mexico, with only about 30 additional species scattered elsewhere around North America. Common to almost all of these originating sites is rocky terrain and quick drainage.

As you may correctly surmise—and as gardeners in many areas have come to appreciate—the general requirement of well-drained soil allows sedums from far-off places to prosper in similar growing conditions. Horticulturists have also been able to "introduce" species of divergent origins and watch for or nurture crosses that may lead to interesting and improved garden varieties, like *Hylotelephium* 'Bertram Anderson', which got its nice plum foliage color from one parent and larger plant size from another.

Hylotelephium spectabile, with its extra-long stamens.

Flower Forms

Arguably many sedums are appreciated primarily for their foliage forms and colors, which have been briefly differentiated above and can be appreciated plant by plant in the plant descriptions themselves, but there are more to sedums than good-looking succulent leaves. As explained elsewhere, botanists acknowledge a plant as a sedum based on flower form. Indeed, the basic characteristics of a sedum flower can be very important to positive identification. Let's take a closer look, then. I'll review the general botanical terms first, then tell you what to look for with sedums.

The male part of most flowers consists of a filament, the thin stem that holds the anther, while the anther is the part that holds the pollen together. A filament and anther together make up the stamen. Most sedums have twice as many stamens as the other flower parts, for example, petals. Sedums also have two-lobed anthers. In some hylotelephiums, stamens extend beyond the petals, giving the flowerheads an almost fuzzy look in their prime. Some hybrids lack stamens altogether and thus do not have pollen to offer pollinators or plant breeders for that matter—they are therefore sterile plants. An example of this phenomenon is *Hylotelephium* 'Herbstfreunde'.

The female organs in flowers are collectively called pistils. A typical pistil consists of one or more carpels; each carpel is made up of a style, a stigma, and the ovary. Carpels are another key flower part to use in identifying sedums. As sedum seeds ripen, the carpels split vertically and—depending on the species—will exhibit one of two distinct reactions. In some, carpels expand on the inside, causing them to arch away from each other toward the petals. Botanically this is called "kyphocarpic." In other sedums, the carpels remain upright, which is called "orthocarpic."

The outer flower parts, that is, sepals and petals, can also aid in positive identification of sedums. Sepals are the leaflike bracts that hold a flowerbud before it opens (think of a rosebud, for instance). In sedums, the number of sepals always equals the number of petals. While most sedum flowers have four to six petals, they can have as few as three and as many as eleven petals per flower. Sepal size is another aspect to use when trying to positively identify sedum species—some have ample sepals (such as *Sedum ternatum*), some are fairly small in size compared to the petals (such as *Petrosedum rupestre*). Lastly, individual sedum flowers are almost always star-shaped (stellate).

Of course, most sedums carry their flowers in groups or clusters, which are then called an inflorescence. An inflorescence that is fairly flat-topped or dome-shaped is called a flowerhead; many hylotelephiums qualify. One that terminates a stem is called a cyme or a panicle. A cyme is a branched cluster having different shapes and densities. A sedum with cyme flowers would be *Phedimus spurius* 'Fuldaglut' (fireglow).

Phedimus kamtschaticus, show-ing a kyphocarpic seedhead with spreading carpels.

Hylotelephium telephium 'Yellow Xenox', showing its orthocarpic seedhead with upright and swollen carpels.

Petrosedum rupestre flowers in bud showing their green sepals, which are relatively small com-pared to the petals they clasp.

Sedum ternatum, with relatively large green sepals next to thin white petals.

Rhodiola rosea f. *arctica* bears its flowers in its leaf axils.

There are, as always, some exceptions to these generalizations. A handful of sedums carry their flowers in the shape of a spike, for instance, *Orostachys boehmeri*.

As for sedum stems, they are almost always round and smooth. There are some exceptions, like species that hold onto old foliage, making their stems look hairy, as in the case of *Sedum multiceps*. Sedum stems can be upright, creeping, or hanging. Like the foliage and often the flowers, sedum stems are colorful or become colorful over the course of a growing season, which undeniably adds to the plants' appeal.

Last but not least, the roots of sedums can also be an identification aid. They can vary from woody to rhizomatous to fibrous. As mentioned above, roots of *Hylotelephium* species are tuberous. In the case of *Rhodiola*, the rhizome grows above the ground with small scalelike leaves at the top of the root.

Subspecies and Hybrids

Sedums as a group are susceptible to diversifying. Certain species, notably *Sedum album*, are especially prolific in their ability to vary and for that reason there are not only many varieties or cultivars but also quite a few subspecies. Subspecies are naturally occurring variations of a species that are tied to a distinct geographical area but may still interbreed

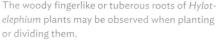

The woody fingerlike or tuberous roots of *Hylotelephium* plants may be observed when planting or dividing them.

Sedum hirsutum with its already small foliage coated in tiny hairs.

with other populations where their distributions overlap. In the case of *S. dasyphyllum* and *S. dasyphyllum* subsp. *glanduliferum*, both are hairy, but the subspecies is more so.

In nurseries, plant catalogs, and even some botanical references, these subtle but real differences may not be recognized, but I believe it is necessary and indeed helpful to use the complete and correct names where applicable. Acknowledging a plant as a subspecies becomes more important, for instance, when it is a parent of a new plant and you wish to understand or trace its unique characteristics.

In the case of hybrids, sometimes the parentage is distinctly from one parent and therefore the new plant looks like a species. This is the case in *Hylotelephium* 'Herbstfreude', which is a hybrid of *H. spectabile* and *H. telephium* but looks so strikingly like *H. spectabile*. In the case of *H. telephium* hybrids, many of the traits come from a subspecies, so the change can be subtle and the plants don't look so very different from the species.

Since many of the best new hybrids on the market today are derived from *Hylotelephium telephium*, I wanted to spend a moment on the major distinctions of its subspecies. *Hylotelephium telephium* subsp. *telephium* and *H. telephium* subsp. *fabaria* have purple to pink-reddish flowers on green foliage on the smaller side of the *H. telephium* plant-size scale. On the other hand, *H. telephium* subsp. *maximum*, with creamy white flowers, and *H. telephium* subsp. *ruprechtii*, usually have red arching stems with glaucous blue-toothed foliage. Many of the initial selections in the trade were just vegetative forms of a special subspecies trait.

Misnamed and Renamed Sedums

Plants can get wrong names attached to them for a variety of reasons. Once people start using a name—right or wrong—it is hard to get them to change. Over the course of writing this book, I have learned the true identity of some of the most commonly misnamed sedums in the horticulture trade. As I discuss and describe these plants here and there in this book, I have corrected these common mistakes and I hope the proper information will "stick" going forward.

One that really surprised me was *Sedum polytrichoides* 'Chocolate Ball', which is almost always listed incorrectly as *S. hakonense*. Other than the fact that both plants are from Japan, I can't quite account for this common error.

A decade ago, while researching the history of *Hylotelephium* 'Herbstfreude' for a patent, I made an interesting, even astonishing, discovery. This is the plant still widely known by its old name and English translation "*Sedum* Autumn Joy." At any rate, I learned that it is a sterile hybrid between two different species. It does not even have male flower parts! The misconception is all the more startling when you consider that much of the time it has been incorrectly listed or sold as *S. spectabile*—now correctly called *Hylotelephium spectabile*—which is one of its parents. (The other parent is *H. telephium*.)

In the American South and West, where many Mexican species of sedum are sold, along with relatives that have similar features, such as *Pachyphytum*, *Pachysedum*, and *Echeveria*, a plant called *Sedum carnicolor* is often sold. This is not an existing species and is therefore invalid. My best guess is that the plant is actually a pachyphytum.

Two Mexican species that get confused are *Sedum adolphii* and *S. nussbaumerianum*. Often it is simply because they are synonyms; *S. adolphii* is the older, correct name. To further complicate the issue, the plant commonly sold as *S. adolphii* is frequently another plant, a hybrid called 'Golden Glow'. 'Golden Glow' has fleshy light green to yellow leaves whose shape resembles *S. adolphii*'s leaf shape. The plant commonly sold as *S. nussbaumerianum* 'Coppertone' is fleshy orange to almost terra-cotta color in the heat of summer; its correct name is *S. adolphii* 'Coppertone'.

Sometimes plants are sold with a newer species name, which by default is invalid since the older name takes precedence. This is the case with *Sedum divergens*, which is sometimes offered under the synonym *S. globosum*. Also, even though *Hylotelephium alboroseum* is much easier to say and spell than its older name *H. erythrostictum*, the older name is correct. *Sedum sediforme* is another one that gets sold by its synonym *S. nicaeense*.

Occasionally a plant just resembles another species and gets labeled as such, and the misnomer sticks. A classic example is the tender *Sedum mexicanum* 'Lemon Ball', which has been sold as the hardy species *Petrosedum rupestre* because the golden foliage strongly resembles *P. rupestre* 'Angelina'. This could be a problem for a gardener who has looked it up prior to shopping and brings home a much more tender plant than he or she expected.

Occasionally a plant just resembles another species and gets labeled as such, and the misnomer sticks.

Originally named *Sedum* 'Amber', this hybrid was transferred to the genus *Hylotelephium* and is now known correctly as *Hylotelephium* 'Amber'.

Wrong names also get applied when similar sedums are not in flower, which is the way the taxonomists tell species apart. For instance, when you compare *Sedum hispanicum* with *S. pallidum* foliage, you will not see much of a difference, but when the plant blooms, you will be able to make a determination: if the flower has pink in it, it's *S. hispanicum*, whereas if it is all white, it's *S. pallidum*.

Taxonomic reorganization that has led to name changes can cause spelling changes, which can be a minor source of confusion. When *Sedum kamtschaticum* was reclassified under the new genus *Phedimus*, *S. kamtschaticum* subsp. *ellacombianun* was changed to its own species; the new name become *Phedimus ellacombeanus*. Notice that the spelling at the end of the species name changed also. That's because of a botanical convention that the word endings should match if possible. It also depends on which part of the plant is being described in Latin, so when the genus was *Sedum*, many of the species ended in *um*, and now that the species is *Phedimus*, many will end in *us*.

Common names can create problems, too. A classic example is *Sedum acre* and *S. sarmentosum*. They share the common name "gold moss," so each can be sold as the other and cause a lot of confusion.

Mislabeled plants among sedums are more common than anyone might like. Much of the time the misnaming took place decades ago and has been perpetuated over time. When I began to research this book, I received *Sedum glaucophyllum* as *S. nevii*, a common mistake. The true *S. nevii* is not actually sold in the trade so this was a relatively easy fix. Same with *S. mocineanum* and *S. hintonii*, both hairy plants and both hailing from Mexico—but *S. hintonii* is not in cultivation, so any plants offered with that name are actually the former plant. Detective work along these lines is not always so neat or easy. For instance, *S. glaucophyllum* is available as two distinct forms in the trade, a darker gray form with red highlights that I propose calling 'Red Frost' and a light green-silver form with the cultivar name of 'Silver Frost' attached to it.

Finally, in the interest of being up-to-date (though still more changes could be ahead, of course), the plants described in this book are listed alphabetically by their most currently accepted names. Synonyms and former names are mentioned in the description. To further help readers find their way, the most commonly used names, old or new, wrong or right, are listed in the index.

150 SEDUMS FOR THE GARDEN

Hylotelephium 'Amber'

Amber stonecrop
SYNONYM *Sedum* 'Amber'

Unlike most hylotelephiums, this hybrid has upright then arching stems that only reach 8–10 inches (20–25 cm) tall. Its stems are darker than the slightly toothed foliage, starting a rose color and deepening in cooler weather. Foliage is mostly glaucous and green gray with some plum highlights by fall. Flowers open in early autumn, with round, 2-inch (5-cm) clusters that have pink petals and deeper pink carpels. My best guess, after looking at the flowers and habit, is that *Hylotelephium ussuriense* is one parent of this hybrid; the flowers are earlier and a bit deeper in hue, but otherwise it is very reminiscent. New in the United States as of 2011, from Skagit Gardens. Deciduous. Patented; propagation is prohibited.

ZONES 4–9

PLANT SIZE 8–10 inches (20–25 cm) tall, 12 inches (30 cm) across.

SOIL dry to well-drained, average

LIGHT full sun

SIMILAR SPECIES AND CULTIVARS *H. sieboldii, H. ussuriense* 'Turkish Delight', *H.* 'Dazzleberry'

ORIGIN bred by Florensis in the Netherlands.

LANDSCAPE AND DESIGN USES The first plants I received arrived in two forms, one with darker foliage and deeper, more ruby flowers and the other with more gray-green foliage and pink flowers. A good one for pots. Try it with silver foliage such as lamb's ears.

Hylotelephium anacampseros

Evergreen orpine
SYNONYM *Sedum anacampseros*

This species could pass for an *Orostachys*, or dunce cap, at certain times, but the flower is completely different. It blooms on buds that developed on last year's stems. The leaves, which become smaller toward the tips, are clustered on the ends of the bare stems. Round and full buds begin gray with plum-pink highlights, then open to five short smoky red petals not much longer than the sepals, arrayed around orange-yellow carpels. It can be shy to flower but is long-lived. Semievergreen. Propagate by cuttings or division.

ZONES 5–9, maybe hardier
PLANT SIZE 5 inches (13 cm) tall, 10 inches (25 cm) across, flower stems 3–4 inches (7.5–10 cm) tall.

SOIL prefers acidic, some moisture; resents limestone
LIGHT full sun to part shade
SIMILAR SPECIES AND CULTIVARS *Hylotelephium anacampseros* f. *majus* is a larger form with more rounded foliage.
ORIGIN Spain, France, and Italy.
LANDSCAPE AND DESIGN USES This is a plant for the rock garden. It is especially nice where it can hang over the edge of a wall or the rim of a container. I have seen it performing nicely in a living wall panel that had an eastern exposure. It would go well with other hanging plants like basket of gold (*Aurinia saxatilis*)

Hylotelephium 'Beach Party'

SYNONYM *Sedum* 'Beach Party'

Its red, almost wine-colored, tipped and edged light olive green foliage keeps this variety interesting from spring to fall. Its compact growth habit is another plus. Once the wide and dense green buds open in early autumn, they begin to touch each other, giving an undulating and full effect like a cloud. Fall flower color is a pleasing light pink with deeper carmine carpels. When open, the flowerheads measure 4 inches (10 cm) wide. Deciduous. Patented; propagation is prohibited.

ZONES 4–9

PLANT SIZE 18 inches (45 cm) tall, 19 inches (48 cm) across.

SOIL dry to well-drained, average

LIGHT full sun

SIMILAR SPECIES AND CULTIVARS *Hylotelephium* 'Dynomite' is new in 2013 from Terra Nova Nurseries; it's a shorter plant but of similar color. *Hylotelephium* 'Pure Joy' is more on the green side and a slightly smaller plant, to 12 inches (30 cm) tall, with lighter pink flowers.

ORIGIN Terra Nova Nurseries, introduced around 2011.

LANDSCAPE AND DESIGN USES This plant's extra-full habit gives a really nice effect in the garden when planted en masse. For a nice contrast with its broad leaves and cloudlike habit, try it with finer foliage like prairie dropseed (*Sporobolus heterolepsis*) or another thin-leaved plant.

Hylotelephium 'Beka'

Autumn Delight stonecrop
SYNONYM *Sedum* 'Beka'

Butter-yellow centers edged in a thin blue-green toothed margin are unique to this variety. Despite the variegation, this plant has good vigor. In late spring and early summer, the variegation is most pronounced; as the season progresses, the distinction fades (though it is always lighter in color than *Hylotelephium* 'Herbstfreude'). Because of the variegation, plants are not as tall. Buds are green, and flowers color up a bit slower. Seed heads turn reddish brown and hold on all winter. Deciduous. Patented; propagation is prohibited.

ZONES 3–9
PLANT SIZE 15–18 inches (38–45 cm) tall, 18 inches (45 cm) across.
SOIL dry to well-drained, average
LIGHT full sun to part shade
SIMILAR SPECIES AND CULTIVARS Autumn Delight is the trademark name and plants are usually sold under this name. *Hylotelephium* 'Herbstfreude' is quite similar; *H.* 'Lajos' has yellow-edged variegation. *H. erythrostictum* 'Mediovariegatum' is also similar, but its leaves may revert to plain green.
ORIGIN found as a sport of *Hylotelephium* 'Lajos' (which has a creamy yellow edge); introduced by the author in 2007.
LANDSCAPE AND DESIGN USES The bright yellow spring foliage makes a good companion in the border, combining easily with most everything. Like any variegated sedum, this one can revert to all-green shoots, which should be removed since they can outgrow the original plant. For extra drama, pair with other yellow-foliage plants such as *Sedum makinoi* 'Ogon' or *Carex elata* 'Bowles Golden'.

Hylotelephium 'Bertram Anderson'

SYNONYM *Sedum* 'Bertram Anderson'

Consistently gray foliage upon emergence is medium-size, measuring around 1 inch (2.5 cm) long and wide. As the season progresses, the foliage turns plum colored. Stems are initially upright, then sprawling, and change from mauve-rose to burgundy red. Loose clusters of flowers bloom at the tips in late summer. The ruby carpels and pink petals are offset nicely by the dark-hued foliage. One of the parents is *Hylotelephium cauticola*. Deciduous. Propagated by division or cuttings. Received the Award of Garden Merit (AGM) from Britain's Royal Horticultural Society in 1993.

ZONES 4–9

PLANT SIZE 6–8 inches (15–20 cm) tall, 12–15 inches (30–38 cm) across.

SOIL dry to well-drained, average

LIGHT full sun

SIMILAR SPECIES AND CULTIVARS Quite a few, among them *Hylotelephium cauticola*, *H. cauticola* 'Lidakense', *H.* 'Dazzleberry', and *H.* 'Ruby Glow'. *Hylotelephium* 'Vera Jameson' is very similar but its coloration is more gray than plum.

ORIGIN garden origin

LANDSCAPE AND DESIGN USES These plants can vary greatly depending on their site. With more sun and leaner soil, the plant grows tighter and neater. Grow it with low silver foliage such as *Veronica incana* 'Pure Silver'.

Hylotelephium 'Birthday Party'

SYNONYM *Sedum* 'Birthday Party'

A real beauty! Fresh green glossy foliage emerges in spring, quickly taking on a red edge. As the season continues, the leaves deepen. Extra-full buds appear gray-green to almost white. Once the flowers open, they tend to overlap. Upon opening, the color is a mid to deep pink; the petals are lighter in color than the carpels, but deepen with age. Deciduous. Patented; propagation is prohibited.

ZONES 4–9

PLANT SIZE 11 inches (28 cm) tall, 20+ inches (50+ cm) across.

SOIL dry to well-drained, average

LIGHT full sun

SIMILAR SPECIES AND CULTIVARS *Hylotelephium* 'Beach Party' has lighter pink flowers but similar foliage, *H*. 'Class Act' has nearly the same color flower but is taller and has darker foliage, and *H*. 'Dynomite' has similar color but is shorter in stature.

ORIGIN Terra Nova Nurseries, introduced in 2010.

LANDSCAPE AND DESIGN USES Due to its smaller size, this one might be better at the front of the border or as a container plant. The brightness of the green foliage makes a nice contrast with other medium-size sedums. Also try it with shorter fall grasses, such as autumn moor grass, *Sesleria autumnalis*.

Hylotelephium 'Blue Pearl'

SYNONYM *Sedum* 'Blue Pearl'

The green-blue foliage is thick, with an almost plasticlike texture. With age, the color improves to a more dusky blue hue. Red stems have almost round leaves that grow close together for a bushy effect. Dark pink flowers begin in late summer. Deciduous.

ZONES 4–9

PLANT SIZE 20 inches (50 cm) tall, 18 inches (45 cm) across.

SOIL well-drained

LIGHT full sun

SIMILAR SPECIES AND CULTIVARS *Hylotelephium* 'Sunset Cloud' is quite similar but its foliage is grayer; *H*. 'Thunderhead' makes substantial plants with darker foliage and purple-pink flowers.

ORIGIN garden, bred by Chris Hansen of Great Garden Plants.

LANDSCAPE AND DESIGN USES The dark blue foliage can go with almost anything. It looks especially ravishing with yellow foliage, such as *Sedum makinoi* 'Ogon', or yellow flowers such as *Coreopsis verticillata* 'Moonbeam' and *Achillea* 'Moonshine'. Add some purple or blue flowers, like *Salvia nemorosa* 'Wesuve', and you have a combination that lasts from spring to fall.

Hylotelephium 'Carl'

SYNONYM *Sedum* 'Carl'

Vigorous, consistent plants are glaucous gray-green with nicely toothed foliage. Upright stems typically have a rose-purple color but are neither dark nor showy, thus never detracting from the flowers. The round flowerheads are reminiscent of *Hylotelephium* 'Herbstfreude' in both size and shape, but the color is a nice, clear bright pink with darker carpels and stamens that are as long as or just barely longer than the petals. Also, this one blooms earlier, in midsummer. Deciduous. Propagation by division.

ZONES 4–9

PLANT SIZE 18–24 inches (45–60 cm) tall and wide.

SOIL dry to well-drained, average

LIGHT full sun

SIMILAR SPECIES AND CULTIVARS *Hylotelephium* 'Abbeydore' and *H. telephium* subsp. *telephium* 'Munstead Dark Red' both have similar foliage and plant size, but have red flowers. Others that are similar include *H.* 'Herbstfreude', *H.* 'Class Act', and *H.* 'Mr. Goodbud'.

ORIGIN found as a seedling of *Hylotelephium spectabile* 'Meteor' at Monksilver Nursery in the United Kingdom.

LANDSCAPE AND DESIGN USES This is a great plant that deserves more appreciation and use. Try pairing it with large blue-leaved grasses like *Panicum virgatum* 'Northwind'.

Hylotelephium cauticola 'Lidakense'

SYNONYM *Sedum cauticola* 'Lidakense'

This short, stout plant spreads by underground stolons, slowly forming spreading mats. Upright then arching stems carry nearly round foliage, glaucous gray with tiny specks of purple; the margins are barely serrated. The opposite foliage is relatively thick for its size at ⅛ inch (3 mm) deep but only 1 inch (2.5 cm) long and wide. Plants bud terminally at the end of summer, and before the flowers start to open, they are cloaked with small bracts just under the developing buds—an intriguing and attractive sight. Early autumn flowers are ruby colored and look handsome against the gray foliage. Deciduous. Easy to increase by division in spring. Received the Award of Garden Merit (AGM) from Britain's Royal Horticultural Society in 2006.

ZONES 4–9
PLANT SIZE 4–6 inches (10–15 cm) tall, 11–12 inches (28–30 cm) across.

SOIL dry to well-drained, average
LIGHT full sun
SIMILAR SPECIES AND CULTIVARS The species, *Hylotelephium cauticola*, is less common and the true form is not easily found. *Hylotelephium cauticola* 'Cola Cola' is touted as more compact. Both *H. cauticola* 'Lidakense' and 'Robustum' may be hybrids of *H. telephium* subsp. *maximum*.
ORIGIN garden origin; species native to Japan.
LANDSCAPE AND DESIGN USES This one is especially drought tolerant and thus always looks fresher at the end of the summer than some of its relatives. It is one of the best sedums for containers. For season-long interest, try pairing it with short-statured *Nepeta* 'Early Bird'; the nepeta's smoky blue flowers look great against the sedum's smoky blue foliage. You could also add some creeping phlox, *Phlox subulata* 'Blue Emerald'.

Hylotelephium 'Cherry Tart'

SYNONYM *Sedum* 'Cherry Tart'

This is a very unique and uniquely beautiful sedum. Thick leaves of plum red are congested on tight clumps. In midsummer, the sprawling stems display blooms in 2- to 3-inch (5- to 7.5-cm) wide clusters. The lightly cup-shaped petals are pink, with a darker stripe down the middle of each one (carpels are ruby red, as are the stamens). Deciduous. Patented; propagation is prohibited.

ZONES 4–9
PLANT SIZE 6 inches (15 cm) tall, 12–15+ inches (30–38+ cm) across.
SOIL dry to well-drained, average
LIGHT full sun
SIMILAR SPECIES AND CULTIVARS *Hylotelephium ussuriense* 'Turkish Delight' has the same foliage color but is a taller plant. *Hylotelephium* 'Bertram Anderson' and *H.* 'Ruby Glow' are comparable.
ORIGIN new in 2012 from breeder Chris Hansen of Great Garden Plants.
LANDSCAPE AND DESIGN USES This medium- to small-size sedum is tight enough that it would be better used as a low edging plant than as a groundcover. Due to its dark leaves, it makes a great contrast for lighter-colored foliage plants. Try it with blue fescue, *Festuca ovina*.

Hylotelephium 'Chocolate Drop'

Chocolate stonecrop
SYNONYM *Sedum* 'Chocolate Drop'

This is one of the most compact red-leaved autumn stonecrops out there. Growth begins deep green but quickly turns dark burgundy red. Scalloped glossy leaves are more congested at the base and then become spaced out as the deeper red stems elongate and start budding. Flowers open in late summer with lighter pink petals and ruby carpels. Deciduous. Patented; propagation is prohibited.

ZONES 4–10
PLANT SIZE 10 inches (25 cm) tall, 14 inches (36 cm) across.
SOIL dry to well-drained, average
LIGHT full sun
SIMILAR SPECIES AND CULTIVARS *Hylotelephium* 'Black Beauty' is quite similar; *H.* Postman's Pride' and *H.* 'Purple Emperor' have lighter purple foliage; *H.* 'Raspberry Truffle' has a lighter color flower and more purple foliage. Overall, I consider this an improved *Hylotelephium* 'Postman's Pride'.
ORIGIN Terra Nova Nurseries, introduced in 2010.
LANDSCAPE AND DESIGN USES The shorter stature makes this an ideal companion in containers. Sharp drainage is important on all these red-leaved ones, since they are prone to rot. Try it alongside pigsqueak, *Bergenia cordifolia*, which has large, shiny green leaves that turn bronze in the fall.

Hylotelephium 'Class Act'

SYNONYM *Sedum* 'Class Act'

This handsome, compact plant is aptly named. It is mostly medium green but has a slight glaucous cast that is more pronounced on the lighter green to rose-colored stems. Light green buds open earlier than most *Hylotelephium*. Both the petals and the carpels of its flowers are nearly ruby pink—the uniformity of the flower color is telltale. Large flowerheads 4+ inches (10+ cm) wide overlap, giving the plant a plush overall look in early autumn. Deciduous. Patented; propagation is prohibited. Received the Award of Garden Merit (AGM) from Britain's Royal Horticultural Society in 2006.

ZONES 4–10

PLANT SIZE 18 inches (45 cm) tall, 24+ inches (60+ cm) across.

SOIL dry to well-drained, average

LIGHT full sun

SIMILAR SPECIES AND CULTIVARS *Hylotelephium* 'Mr. Goodbud' is the same size, but has darker purple flowers. *Hylotelephium telephium* subsp. *telephium* 'Munstead Dark Red' has a similar habit and red flowers.

ORIGIN Terra Nova Nurseries, introduced in 2005.

LANDSCAPE AND DESIGN USES An easy plant to place in a border, thanks to its fullness and medium-to-large size for an upright sedum. I recommend planting it among shorter pink coneflowers, for instance, *Echinacea* 'Satin Nights'.

Hylotelephium 'Cloud Walker'

SYNONYM *Sedum* 'Cloud Walker'

'Cloud Walker' offers quite a color show. At spring emergence, the foliage is deep green but quickly takes on some purple-red highlights, eventually becoming half purple and dark green. Leaf edges are unevenly toothed. Stems are mostly purple, and topped with creamy greenish white buds. The flower color is unique, ultimately more wine purple than pink (petals are white with pink, while purple carpels bleed color into the petals, turning them darker). Branched flowerheads can be extra-large, reaching 6 inches (15 cm) or more across. Deciduous. Patented; propagation is prohibited.

ZONES 4–9
PLANT SIZE 18–24+ inches (45–60+ cm) tall and wide.
SOIL dry to well-drained, average
LIGHT full sun
SIMILAR SPECIES AND CULTIVARS *Hylotelephium* 'Maestro'—the main difference is that this flower color is more bicolor pink; *H.* 'Matrona' is also similar.
ORIGIN Terra Nova Nurseries, introduced around 2005.
LANDSCAPE AND DESIGN USES This plant is prone to disease and therefore can be short-lived unless placed in an ideal spot (highly drained sites or very lean soil). Thanks to its purple hues, it combines easier than some of its red cousins. It looks great with other purples, for example, *Vernonia lettermanii* or *Origanum laevigatum* 'Herrenhausen'. Its fall coloration fits in naturally with all the other fall colors.

Hylotelephium 'Dazzleberry'

SYNONYM *Sedum* 'Dazzleberry'

Chalky purple foliage adorns sprawling leafy red stems. The leaves are almost round and, at times, slightly toothed (dentate). Late summer brings flower stems as well as a new spurt of growth in the plant's center. Buds begin pink and gray-striped, soon opening to heavy heads of carmine pink with white at the base of the petals. Very vigorous. Deciduous. Patented; propagation is prohibited.

ZONES 4–9
PLANT SIZE 8–10 inches (20–25 cm) tall, 18+ inches (45+ cm) across.
SOIL dry to well-drained, average
LIGHT full sun
SIMILAR SPECIES AND CULTIVARS *Hylotelephium* 'Sunset Cloud'; *H.* 'Bertram Anderson' has a looser habit and smaller foliage.
ORIGIN bred by Chris Hansen of Great Garden Plants, introduced in 2012.
LANDSCAPE AND DESIGN USES These sprawling plants are ideal for the edge of a raised bed where growth can hang over the edge. If planted close enough, they will weave into other low to medium plants nearby. Planting it with short grasses like *Sporobolus heterolepsis* 'Tara' would be pleasing.

Hylotelephium erythrostictum 'Frosty Morn'

SYNONYM *Sedum erythrostictum* 'Frosty Morn'

A striking variegated sedum, beloved by many. Though the thin yet wide foliage is reminiscent of *Hylotelephium spectabile*, the markings are dramatic: light green centers edged somewhat irregularly in a white band, typically less than ⅛ inch (3 mm) wide. By late summer, upright growth terminates in a cluster of bracts surrounding the flowerbuds. These bracts can actually be as showy as, or showier than, the actual flowers. The flowers appear light pink but are made up of white petals surrounding light pink carpels, which together form the ovary. Deciduous. Propagate by division.

ZONES 4–9
PLANT SIZE 28 inches (70 cm) tall, 18–24 inches (45–60 cm) across.
SOIL well-drained, average
LIGHT full sun
SIMILAR SPECIES AND CULTIVARS *Hylotelephium spectabile* 'Pink Chablis' has similar foliage.
ORIGIN garden origin; introduced in Japan by Gotemba Nursery and brought to the United States by Barry Yinger, who gave it to Hines Nursery to introduce.
LANDSCAPE AND DESIGN USES Grow this plant with ornamental grasses that echo the white variegation, such as *Calamagrostis* ×*acutiflora* 'Overdam' or *Miscanthus sinensis* 'Morning Light'. Remove any all-green shoots as close to the ground as possible (if left on the plant, they can outgrow the more desirable variegated parts)

Hylotelephium erythrostictum 'Mediovariegatum'

Clown stonecrop

SYNONYM *Hylotelephium alboroseum, Sedum erythrostictum* f. *mediovariegatum*

A showy, vigorous two-tone choice with green-edged, gold-centered glaucous foliage. Crowded rosettes emerge in early spring, followed by upright (almost vertical) growth cloaked with leaves in opposite pairs. In late summer, plants begin to show buds and leafy bracts, almost smothering the foliage and practically holding the flowers. While the overall look of the distinctive blooms is soft pink, they too are two-tone, with creamy white petals surrounding soft pink carpels. Deciduous. Propagate by division.

ZONES 4–9

PLANT SIZE 24 inches (60 cm) tall, 15–18 inches (38–45 cm) across.

SOIL well-drained, average

LIGHT full sun to partial shade

SIMILAR SPECIES AND CULTIVARS Sometimes sold as 'Clown' or 'The Clown'. Despite being very old and not overly common, this is one of the showiest sedums in spring; the drawback is that you may need to constantly rouge out green reversions. *Hylotelephium erythrostictum* 'Lemonade' is an all-gold form. The green form is not as desirable, as I have never seen it sold. *Hylotelephium telephioides*, which may be a parent (along with *H. viridescens*), is a North American native that also has white or pinkish flowers.

ORIGIN garden origin

LANDSCAPE AND DESIGN USES Due to the soft yellow coloration of the foliage, it is used effortlessly in the spring border when it is showiest. Splendid with grasses like *Calamagrostis* 'Karl Foerster' or 'Eldorado' that will grow up past it later when the yellow foliage gets covered with flowerbuds.

Hylotelephium ewersii subsp. *homophyllum* 'Rosenteppich'

Rose carpet stonecrop

SYNONYM *Sedum ewersii* subsp. *homophyllum* 'Rosenteppich'

Crawling and sprawling, permanently woody stems carry silver green, glaucous, fingernail-shaped foliage with clasping bases. Late in the season, bright pink 1- to 2-inch (2- to 5-cm) wide flower clusters stand on short stems only 3–4 inches (7.5–10 cm) tall. In winter, the plants look like a tangled mass of brown stems with buttons of foliage persisting at the tips. Semievergreen. Increases easily from divisions, but cuttings work too. I've also raised this plant from seed, and it seems to come true.

ZONES 4–9

PLANT SIZE 4 inches (10 cm) tall, up to 6 inches (15 cm) across.

SOIL rocky, dry to well-drained, average

LIGHT full sun

SIMILAR SPECIES AND CULTIVARS Close in size and habit to *Hylotelephium pluricaule*. Plants that are similar but larger and sport larger foliage include the species, *Hylotelephium ewersii* (sometimes incorrectly called *H. sieboldii* var. *minor*), *H. cauticola*, and *H. sieboldii*. *Hylotelephium ewersii* var. *cyclophyllum* also has larger foliage that develops striking rose pink highlights in fall. *Hylotelephium ewersii* var. *homophyllum* is also a smaller form, but otherwise quite similar. *Hylotelephium* 'Lime Zinger' grows 6 inches (15 cm) tall and 18 inches (45 cm) wide, with fresh gray-green foliage that looks nice all season long; its leaves are similar in size to the larger species.

ORIGIN garden origin

LANDSCAPE AND DESIGN USES Makes a nice contrasting pair with the larger species, *Hylotelephium ewersii*, in the garden or in a container. In any event, due to its sprawling nature, it makes a fine choice for container edges or stone wall edges, anywhere it can hang over.

Hylotelephium 'Herbstfreude'

Autumn Joy
SYNONYM *Sedum* 'Autumn Joy', *S.* 'Herbstfreude'

With larger and more gray-green, more toothed foliage than many of it relatives, this plant is hard to miss in the landscape. Emerging plants grow fast in spring, becoming dome shaped mounds of gray-green. Vigorous plants begin to bud in mid-summer, resembling broccoli plants with a creamy green color. Stems are lighter green with no hint of red. As the season progresses, substantial heads usually 4 inches (10 cm) wide or larger begin to open on the blushing pink side but deepen with cool nights until they become brick red. The sterile flowers lack male parts (stamens and anthers). As winter arrives, the leaves eventually drop off, leaving cinnamon-colored stems and seed heads that stand up for months. Deciduous. Easy to propagate from division or cuttings. Received the Award of Garden Merit (AGM) from Britain's Royal Horticultural Society in 1993.

ZONES 3–9
PLANT SIZE 24 inches (60 cm) tall and at least that much across.
SOIL dry to well-drained, average
LIGHT full sun to part shade
SIMILAR SPECIES AND CULTIVARS The very similar-looking *Hylotelephium* 'Autumn Fire' is said to be more compact and does not require pinching or staking. I have not detected a difference, but I have not grown them side by side. *Hylotelephium* 'Indian Chief' is said to be earlier-blooming by a couple weeks, with larger heads (it may be a synonym). *Hylotelephium* 'T-Rex', said to be a sport, has distinctly more pronounced teeth on the edge of the leaves. *Hylotelephium* 'Beka' has gold-centered foliage; *Hylotelephium* 'Lajos' leaves edged creamy yellow. *Hylotelephium* 'Elsie's Gold' is a sport with yellow variegated edges. *Hylotelephium* 'Jaws' is yet another sport, with a wavy-toothed margin to the edge of its foliage. *Hylotelephium* 'White Tooth Shark' has a thin white edge, but reverts back to the original form easily.
ORIGIN a sterile hybrid between *Hylotelephium telephium* and *H. spectabile*, introduced in 1952 by Arends Nursery.
LANDSCAPE AND DESIGN USES This is easily one of the most common and most popular sedums, and for good reason. It has interest in the garden all year. It is very attractive to bees, butterflies, and other pollinators for its nectar. Upon a close sniff plants are pleasingly fragrant (but watch out for the bees). When the buds are greenish white, this plant complements the summer-blooming white hydrangeas like 'Annabelle'. One of my favorite fall combinations pairs *H.* 'Herbstfreude' with *Eragrostis spectabilis*.

Hylotelephium 'Lajos'

Autumn Charm stonecrop
SYNONYM *Sedum* 'Lajos'

For a variegated sedum, this one is impressively stable, which means that you don't have to be constantly vigilant about reverted, all-green leaves. The yellow edge is usually ¼ inch (6 mm) wide, with gray-green centers. The soft yellow edges fade to creamy white. The notched edges are more pronounced at the tips. Its creamy white buds, which debut in late summer, make a showier statement than the evolving pink to red flowerheads that follow in early autumn. Deciduous. Patented; propagation is prohibited.

ZONES 4–9
PLANT SIZE 15–20 inches (38–50 cm) tall, 18+ inches (45+ cm) across.
SOIL dry to well-drained, average
LIGHT full sun to light shade (seems to really benefit from some light shade)

SIMILAR SPECIES AND CULTIVARS Autumn Charm is its trade name, and most nurseries sell it by this name. For nearly the reverse variegation pattern—yellow-centered foliage with a thin green edge—try *Hylotelephium* 'Beka'. For a similar variegation pattern, but more on the yellow side, seek out *H.* 'Elsie's Gold' (this one also holds its yellow color in the leaf later in the season). Finally, for similar foliage but a more compact habit, grow *H.* 'Frosted Fire'.
ORIGIN garden origin, found by the author; introduced in 2006.
LANDSCAPE AND DESIGN USES Its biggest selling point besides the handsome variegation is that it is a sport of 'Herbstfreude'. It looks great on its own or massed, but combines easily with sedums like *Phedimus spurius* 'Tricolor'. It also goes really well with red-leaved heucheras, such as *Heuchera* 'Plum Pudding'.

Hylotelephium 'Maestro'

SYNONYM *Sedum* 'Maestro'

An upright habit with sturdy stems are the main assets of this sport of *Hylotelephium* 'Matrona'. The plants don't splay open in the middle, even when blooming—another real plus! Thus garden performance is more reliable. Also, 'Maestro' has more substantial flowerheads that reach 4 inches (10 cm) or more wide, with a deeper pink color than 'Matrona'. Blooms in summer. Foliage is thick and can turn from dusky gray and green to deeper green or even purple, depending on the exposure. Stem color starts rose, then becomes maroon for most of the season. Deciduous. Patented; propagation is prohibited.

ZONES 4–9
PLANT SIZE 24–30 inches (60–76 cm) tall, 24+ inches (60+ cm) across.
SOIL dry to well-drained, average
LIGHT full sun
SIMILAR SPECIES AND CULTIVARS *Hylotelephium* 'Crazy Ruffles'; *H.* 'Matrona' has lighter foliage and flower color.
ORIGIN found as a sport of *Hylotelephium* 'Matrona' by Gary Trucks of Amber Waves Gardens.
LANDSCAPE AND DESIGN USES Bees and butterflies are strongly attracted to this plant. For a charming picture, grow it in the company of an ornamental grass like dwarf fountain grass, *Pennisetum alopecuroides* 'Ginger Love'.

Hylotelephium 'Matrona'

SYNONYM *Sedum* 'Matrona'

A substantial, robust plant of mostly gray-green, larger foliage with some purple highlights, and medium-size flower clusters that don't always overlap. Buds typically develop in summer and are creamy white; flowers are on the light pink side with white petals tipped pink. Stems are thick, upright, and purple. Foliage is extra thick and can measure 3 inches (7.5 cm) long by 2 inches (5 cm) wide. Deciduous. Propagate by division in spring. This plant won the ISU Award (International Stauden Union, a perennial grower group in Europe) in 2000, and the Award of Garden Merit (AGM) from Britain's Royal Horticultural Society in 1993.

ZONES 3–9

PLANT SIZE 24–32 inches (60–80 cm) tall, 18+ inches (45+ cm) across.

SOIL dry to well-drained, average

LIGHT full sun

SIMILAR SPECIES AND CULTIVARS Both *Hylotelephium* 'Aquarel' and *H.* 'Joyce Henderson' look almost the same, with less red to their gray-green foliage; they are also more resistant to disease. *Hylotelephium* 'Black Jack' (European Community synonym 'Dark Jack'), introduced by Walters Gardens in Michigan, is a darker sport that is relatively unstable (unfortunately it reverts back to green). *Hylotelephium* 'Crazy Ruffles' has a ruffled edge to its foliage; this one is an improved form with deeper foliage and flower color.

ORIGIN discovered in 1986 by Ewald Hugin in Germany, as a seedling of *Hylotelephium telephium* subsp. *maximum* 'Atropurpureum' (puzzlingly, 'Herbstfreude' was named as the pollen parent, which is impossible because it does not produce pollen).

LANDSCAPE AND DESIGN USES Owing to its husky size and fullness, it looks terrific among many of the taller fall plants, notably ornamental grasses and asters.

Hylotelephium 'Mr. Goodbud'

SYNONYM *Sedum* 'Mr. Goodbud'

As the name suggests, this plant has striking buds. They are denser than most, though gray in color. The resulting flowers reliably reach 4 inches (10 cm) or more in diameter and tend to overlap when in bloom; they generally start in late summer and continue into early autumn. They start as a bicolor pink with lighter pointed petals but deepen to match the color of the carpels, becoming a raspberry pink. The robust plants are leafy and green, though the upper leaves can get red edges by the time they bloom, which makes a nice contrast (lower foliage remains mostly green). Stems are rose with a slight glaucous cast. Deciduous. Patented; propagation is prohibited. Received the Award of Garden Merit (AGM) from Britain's Royal Horticultural Society in 2006.

ZONES 4–10
PLANT SIZE 15–18 inches (38–45 cm) tall, 15+ inches (38+ cm) across.
SOIL dry to well-drained, average
LIGHT full sun
SIMILAR SPECIES AND CULTIVARS *Hylotelephium* 'Class Act'; *H.* 'Thunderhead' has a similar color but is about twice as big.
ORIGIN Terra Nova Nurseries, introduced in 2006; a cross of *Hylotelephium spectabile* 'Brilliant' and an unnamed *H. telephium*.
LANDSCAPE AND DESIGN USES The greener foliage on this plant is refreshing among the many gray-green forms; thus it makes a fine contrast among other tall autumn stonecrops. Perfect with fall grasses like *Miscanthus sinensis* 'Malepartus'.

Hylotelephium 'Plum Perfection'

Plum stonecrop
SYNONYM *Sedum* 'Plum Perfection'

The tips of this plant start out greenish but soon turn gray and then mostly plum-colored except for the base of the leaf. The narrow serrated leaves measure between ¼ and ⅜ inch (6–9 mm) wide by 1 inch (2.5 cm) or more long. Strongly clumping plants with arching relaxed stems start to bloom in late summer. Buds are creamy white tipped pink and measure 3 inches (7.5 cm) wide. Once open, the flowers are also creamy white with pink tips. Deciduous. Patented; propagation is prohibited.

ZONES 4–9
PLANT SIZE 6–8 inches (15–20 cm) tall, 8–12+ inches (20–30+ cm) across.
SOIL dry to well-drained, average
LIGHT full sun to part shade
SIMILAR SPECIES AND CULTIVARS *Hylotelephium* 'Bertram Anderson' has similar foliage color but a looser, more open habit and larger leaves.
ORIGIN developed by the author, introduced in 2010.
LANDSCAPE AND DESIGN USES This medium-size sedum is very drought tolerant, so it makes an ideal rock- or gravel garden plant. Such settings also allow its dark foliage to play off light-colored stones. I have it planted near golden-leaved *Sedum sexangulare* 'Golddigger' and the combination is gorgeous. Or, if you wish, tuck this one into a pot; its sprawling stems make it a good filler among plants of other sizes, while its plum hue complements different foliage colors in its companions.

Hylotelephium pluricaule var. *ezawe*

SYNONYM *Sedum pluricaule* var. *ezawe*

Even though it is one of the smallest hylotelephiums, this plant has substance, thanks to its fat but tiny opposite leaves with short internodes on lax stems. Blue-gray entire foliage takes on a glaucous pink cast as the season progresses. Shy to bloom—flowers are rare. Evergreen (though it does lose some lower foliage over the winter). Division is the best way to propagate.

ZONES 4–9

PLANT SIZE 1–2 inches (2.5–5 cm) tall, 12–15 inches (30–38 cm) across.

SOIL rocky, dry to well-drained

LIGHT full sun

SIMILAR SPECIES AND CULTIVARS *Hylotelephium ewersii* subsp. *homophyllum* 'Rosenteppich' (rose carpet) is a different color but has a similar leaf size and texture.

ORIGIN Eastern Siberia.

LANDSCAPE AND DESIGN USES The low sprawling nature and the blue to pink miniature foliage make this an ideal candidate for the edge of a wall or walkway, which it will soften. I have grown it in a raised bed at the edge of rock wall, where it thrives. I have also seen it used effectively in a crevice garden.

Hylotelephium populifolium

SYNONYM *Sedum populifolium*

It's easy to identify this unique sedum. It has cinnamon brown, woody stems, and flat-toothed foliage that is a grooved down the middle (also, each leaf is borne on a stalk, a feature seen only on this species). Habit is strongly clumping, with upright to arching stems. In late summer, flowers appear in terminal clusters, usually 2–3 inches (5–7.5 cm) wide. They give an overall pink impression, though a closer look reveals that the petals are white and the carpels are pink. Deciduous. Propagate by spring cuttings or seed.

ZONES 3–9
PLANT SIZE 10–13 inches (25–33 cm) tall, 23 inches (58 cm) across.
SOIL dry to well-drained, average
LIGHT full sun to part shade
SIMILAR SPECIES AND CULTIVARS *Hylotelephium populifolium* 'Maurie's Form' has more pointed foliage.
ORIGIN Siberia near Lake Baikal.
LANDSCAPE AND DESIGN USES This resilient and long-lived deciduous species deserves more use. First and foremost, it is shade-tolerant and secondly, it is quite cold-hardy (which isn't true of plenty of its relatives). For me, it makes an ideal container plant solo or in combination, but it is also perfectly happy outside in the garden. Because of its notched foliage, it would look good with *Hylotelephium* 'Thundercloud'.

Hylotelephium 'Pure Joy'

SYNONYM *Sedum* 'Pure Joy'

This one's full, mounding habit with blooms reaching all the way down to the ground makes a stunning effect. Silver to gray foliage emerges in tight domes, eventually becoming greener; the leaves are toothed and over 2 inches (5 cm) long by 1 inch (2.5 cm) wide. By midsummer, plants begin to branch heavily and develop greenish white buds. In early autumn, when the bicolor pink flowers open, the entire plant looks like one massive ball of flowers (see photo on page 203). Individual flowers start out on the light side but deepen with time and cool nights. The tips of the petals are pink fading to white at the base. Carpels and stamens eventually turn magenta. Deciduous. Patented; propagation is prohibited.

ZONES 4–9
PLANT SIZE 12+ inches (30+ cm) tall, 16–20+ inches (41–50+ cm) across.
SOIL dry to well-drained, average
LIGHT full sun
SIMILAR SPECIES AND CULTIVARS *Hylotelephium* 'Soft Cloud' grows a few inches taller and has light pink flower clusters separated from each other (like most hylotelephiums); *H.* 'Thundercloud' also has separate flower clusters, in white. This plant's foliage is reminiscent of *H. sieboldii*, which sets it apart from some of its relatives.
ORIGIN developed by the author, introduced in 2011.
LANDSCAPE AND DESIGN USES This plant works as a specimen container plant, just like a fall mum would be used. A successful combination I've enjoyed growing places it adjacent to *Schyzachrium scoparium* 'Jazz' and *Penstemon digitalis* 'Pocahontas'.

Hylotelephium 'Ruby Glow'

Ruby Glow sedum
SYNONYM *Sedum* 'Ruby Glow',
S. 'Rosey Glow', *S.* 'Robustum'

Here is a trouble-free, spreading plant with bright, pretty flowers. Its pleasing, glaucous foliage is more green than gray, and more round than long. The foliage darkens in sun. The relaxed stems start upright but then sprawl. Stem color is usually purple and glaucous as well. Buds are mauve at the tips and gray at the base, and can have some bracts (like one of the parents, *Hylotelephium cauticola*). Once fully open in late summer, however, the flowers are quite uniformly vivid pink in color from petal to carpel to stamen. They are borne in loose clusters at the tips and big enough to touch, almost covering the plant. Deciduous. Propagated by division in spring.

ZONES 4–9
PLANT SIZE 10 inches (25 cm) tall, 12–15+ inches (30–38+ cm) across.
SOIL dry to well-drained, average
LIGHT full sun
SIMILAR SPECIES AND CULTIVARS *Hylotelephium* 'Amber' has similar coloration, but a bushier, upright habit. *Hylotelephium* 'Bertram Anderson' has darker foliage color. *Hylotelephium cauticola* has similar color and a tighter mounding habit. *Hylotelephium cauticola* 'Lidakense' and *H.* 'Dazzleberry' have fuller habits and darker foliage, while *H.* 'Vera Jameson' is quite similar but has finer foliage.
ORIGIN raised by German hybridizer Georg Arends and sent to Alan Bloom in England in about 1952; it is a true intermediate of the parents, which are said to be *Hylotelephium cauticola* and *H. telephium*.
LANDSCAPE AND DESIGN USES This one is never aggressive, always playing nice with its neighbors. It seems only to complement, never clash with other plants. Try it with bright yellow Missouri primrose (*Oenothera macrocarpa*).

Hylotelephium sieboldii

October daphne
SYNONYM *Sedum sieboldii*

This species has a lot going for it: great foliage color when it emerges, excellent drought tolerance over the summer months, and a vivid fall show that involves bright leaves and late-appearing flowers, when most other perennials are done for the year. It would be hard to confuse this sedum with any others, due to its distinctive ternate (arranged in threes) foliage with dull lobed teeth on the top edges. It starts out glaucous blue-green, almost silver. Later in the season, the leaf edges typically turn red, especially in a full-sun exposure—the result practically resembles a whole new plant. By fall, the leaves change again, becoming bright orange-red. The plant actually gets its common name from its midautumn flowers; these are two-tone, midpink with darker carpels. Deciduous. Division or spring cuttings are the best ways to propagate it.

ZONES 4–9
PLANT SIZE 10 inches (25 cm) tall, 12+ inches (30+ cm) across.

SOIL dry to well-drained, average
LIGHT full sun to part shade
SIMILAR SPECIES AND CULTIVARS *Hylotelephium sieboldii* 'October Daphne' may be a true selection that has thinner green foliage and a pronounced red edge after spring. *Hylotelephium* 'Lime Zinger' only grows 6 inches (15 cm) tall and 18 inches (45 cm) across, but at times its foliage takes on a similar red edge (however, its foliage is otherwise more like *H. ewersii*).
ORIGIN Japan
LANDSCAPE AND DESIGN USES This plant tends to look better with some light shade. Sharp drainage is also important. I have planted it at the edge of a rock wall where it can hang over the edge and show off its arching stems. This species may even be grown inside in a pot, but will still go dormant for the winter. Try planting it with any silver-leaved *Heuchera*, for example, 'Silver Scrolls'.

Hylotelephium sieboldii f. variegatum

Variegated October daphne
SYNONYM *Sedum sieboldii* f. *variegatum*

By midseason, this variegated version gains a wonderful tricolor appearance. It is typically smaller than the solid blue-green species and not quite as vigorous. The variegation is hard to see on the emerging foliage in spring and does not always have a regular shape. In fact, unlike most variegated sedums, it does not have a distinct line between the creamy yellow centers and the blue-green sides. Once plants mature, the foliage starts to take on a plum-red edge and the stems darken as well. Plants bloom in midautumn; note that its pink flowers are lighter in hue compared to the species. Deciduous. Propagate by division or cuttings.

ZONES 4–9
PLANT SIZE 6–8 inches (15–20 cm) tall and at least that much across.
SOIL dry to well-drained, average
LIGHT full sun to part shade
SIMILAR SPECIES AND CULTIVARS nothing very similar
ORIGIN garden origin
LANDSCAPE AND DESIGN USES I recommend this form as a specimen in a rock garden or along the edge of a raised bed or a wall. It also makes a nice container plant, either alone or in combination with others. Solid blue-green shoots are relatively common; remove them, or they'll outgrow the variegated parts.

Hylotelephium 'Soft Cloud'

SYNONYM *Sedum* 'Soft Cloud'

These plants are very vigorous and can have over a hundred growth tips on one mature clump! Upon emergence in spring, they make domes that slowly expand and open up until the buds develop in early autumn. Foliage is light gray-green, strongly toothed, and almost whorled around the glaucous stems. The inflorescences can reach 3–4 inches (7.5–10 cm) wide and tend to have some space between them. Flower color is bicolor, blush pink and midpink, which deepen only slightly with cooler weather. Deciduous.

ZONES 4–9

PLANT SIZE 15 inches (38 cm) tall, 24+ inches (60+ cm) across.

SOIL dry to well-drained, average

LIGHT full sun

SIMILAR SPECIES AND CULTIVARS *Hylotelephium spectabile* 'Crystal Pink', at 16 inches (41 cm) tall and wide, and *H.* 'Pure Joy' at only 12 inches (30 cm) tall, are similar. Also similar but shorter is *H.* 'Thundercloud', with white flowers that age to pink. *Hylotelephium spectabile* 'K's Little Joy' is new, an even shorter one with smaller foliage (which suggests that it may be a hybrid).

ORIGIN developed by the author, introduced in 2012.

LANDSCAPE AND DESIGN USES In spring, the round clumps are quite a sight. Because they are so architectural, you could use them in rigid or formal designs such as a deciduous hedge, knot garden, or formal border. Would also look excellent with short- to medium-size ornamental grasses like little bluestem (*Schyzachrium scoparium*) or autumn moor grass (*Molinia autumnalis*).

Hylotelephium spectabile 'Brilliant'

SYNONYM *Sedum spectabile* 'Brilliant'

This plant is a star. It is one of the most durable, trouble-free border plants in the genus and justly popular. The leaves are light green, more glaucous on the back side, with stems of nearly the same color. Leaf size is usually 1+ inch (2.5+ cm) wide and up to 2 inches (5 cm) long, with slightly toothed edges. There is no pattern to leaves on the stem; they vary from opposite to whorled to alternate. The flowers are spectacular; typically all parts are the same color, light pink, with exaggerated stamens that reach out past the end of the petals. The clusters form a flat dome, color up in early autumn, and reach 3–4 inches (7.5–10 cm) wide. Its resistance to fungal diseases, especially *Rhizoctonia*, is a real plus. Deciduous. Propagate by division. Received the Award of Garden Merit (AGM) from Britain's Royal Horticultural Society in 1993.

ZONES 4–9

PLANT SIZE 18–24 inches (45–60 cm) tall and at least that much across.

SOIL well-drained, average

LIGHT full sun to part shade

SIMILAR SPECIES AND CULTIVARS *Hylotelephium spectabile* 'Crystal Pink' and H. 'Soft Cloud' are both shorter. *Hylotelephium spectabile* 'Rosenteller' has mid to light pink flowers.

ORIGIN garden origin; species from Northern China and Korea.

LANDSCAPE AND DESIGN USES 'Brilliant' is a heavy feeder, which is why it is typically light green in color. Its nectar attracts many pollinators, including bees and butterflies. In the trade, flower color may vary from light pink to midpink; while I suppose soil fertility could be a factor, I've always equated the light pink form with this variety. It complements the darker pink *Hylotelephium* 'Herbstfreude' (see photo).

Hylotelephium spectabile 'Brilliant' (foreground)

Hylotelephium spectabile 'Carmen'

SYNONYM *Sedum spectabile* 'Carmen'

Of the cultivars selected for darker, carmine pink flowers, this is one of the best. Glaucous gray-green foliage can be light-colored depending on your site and soil fertility. Rubbery-textured foliage alternates on the similarly colored stems. Substantial flowerheads reach 4 inches (10 cm) or more in diameter and are made up of hundreds of individual dark pink flowers with the telltale stamens extending past the petals. Deciduous. Propagate by division.

ZONES 4–9

PLANT SIZE 18–24 inches (45–60 cm) tall and at least that much across.

SOIL dry to well-drained, average

LIGHT full sun to part shade

SIMILAR SPECIES AND CULTIVARS There other dark pink forms that are probably all fairly similar, notably *Hylotelephium spectabile* 'Lisa' and 'Meteor', both at 24 inches (60 cm) tall, and 'Pink Fairy' and 'Septemberglut', both at 27 inches (68 cm) tall.

ORIGIN garden origin

LANDSCAPE AND DESIGN USES I love the deep color and substantial height of this plant; there are few plants it does not play off of nicely in the fall garden. It is especially beautiful with fall-blooming asters. Its main drawback is that the seedheads do not hold up to winter weather and snow like *Hylotelephium* 'Herbstfreude'.

Hylotelephium spectabile 'Neon'

SYNONYM *Sedum spectabile* 'Neon'

Short stature and rich, pink carmine flowers are the prominent features of this cultivar. Due to its shorter growth habit, its leaves maybe slightly more congested and individually, on the smaller side. Deciduous. Easy to propagate from division in early spring.

ZONES 4–9

PLANT SIZE 15–18 inches (38–45 cm) tall, 18+ inches (45+ cm) across.

SOIL dry to well-drained, average

LIGHT full sun

SIMILAR SPECIES AND CULTIVARS There are quite a few similar, relatively shorter selections of *Hylotelephium spectabile*. 'Hot Stuff' is in the range of 16 inches (41 cm) tall and wide. 'Mini Joy' is another short form that has light pink flowers like the species. 'Pink Bomb' has large flowerheads of deep pink. An intriguing choice would be 'Pippin Purple', with thinner petals that are creamy white on the backside, giving the whole flowerhead a silvered look; it gets around 15 inches (38 cm) tall. Shortest yet would be 'Pizzazz', a newer one from Plant Haven, growing only 12 inches (30 cm) tall and wide but sporting an extra-large dome-shaped flowerhead of deep pink. Finally, two others I have heard good things about but never seen for sale in the United States are 'Steven Ward' (it is available in England) and 'Abendrot', the latter 12 inches (30 cm) tall with purple-pink flowers.

ORIGIN garden origin

LANDSCAPE AND DESIGN USES Since this is a shorter variety, grow it with similar-size plants. Short ornamental grasses like *Sporobolus heterolepsis* 'Tara' make a perfect match.

Hylotelephium spectabile 'Pink Chablis'

SYNONYM *Sedum spectabile* 'Pink Chablis'

An interesting plant! While there are others with variegated foliage, so far this is the only one derived from *Hylotelephium spectabile*. Its handszome gray-green glaucous foliage has a relatively thin creamy white edge. In the bud stage, 'Pink Chablis' appears entirely white and almost looks like it is already in flower. However, once open, flowers are bicolor with white petals (look closely—the extra-long stamens are white). Deciduous. Patented; propagation is prohibited.

ZONES 4–8

PLANT SIZE 18–20 inches (45–50 cm) tall, 18–24 inches (45–60 cm) across.

SOIL dry to well-drained, average

LIGHT full sun

SIMILAR SPECIES AND CULTIVARS *Hylotelephium erythrostictum* 'Frosty Morn' has similar white-edged foliage, but is taller and its flower color is mostly white with hints of pink.

ORIGIN garden origin, found by Christopher Howe; introduced in 1995 by Hortech Nursery in Michigan.

LANDSCAPE AND DESIGN USES The thin white variegation brightens this plant and makes it easy to blend into garden beds. I like it with other white-variegated or silver foliage like *Phedimus spurius* 'Tricolor' at its base or silver-leaved chives such as *Allium schoenoprasum* 'Forescate'. Be forewarned that this sedum can revert to vigorous all-green shoots; remove these close to the ground, since if left on the plant, they will outgrow the variegated parts. Plants can also sport all-white shoots that don't fully develop or flower—these, too, should be removed (if you don't, the tips may eventually turn brown and rot).

Hylotelephium spectabile 'Stardust'

SYNONYM *Sedum spectabile* 'Stardust'

A very fine white-flowered form that is trouble-free and long-lived. Tight but substantial clusters reach 3–4 inches (7.5–10 cm) tall and contain hundreds of individual flowers. Before they open, the buds have a slight greenish cast to them. Once fully open, however, the flowers are entirely white, including the ovary and extra-long stamens that extend past the petals. Deciduous. Propagate by division.

ZONES 4–9

PLANT SIZE 18–24 inches tall (45–60 cm) and at least that much across.

SOIL dry to well-drained, average

LIGHT full sun to part shade

SIMILAR SPECIES AND CULTIVARS *Hylotelephium spectabile* 'Snow Queen' is the other commonly sold white form. The blooms of 'Iceberg' are more creamy white.

ORIGIN garden origin

LANDSCAPE AND DESIGN USES Grow it with similar-size, related border sedums; the contrast this form provides is easy on the eyes. Its white flowers also combine well with fall grasses such as *Panicum virgatum* 'Shenandoah'. Please note that, like all the white-flowered forms, there tends to be some reversion to pink; simply remove the unwanted stems as close to the ground as possible.

Hylotelephium 'Sunset Cloud'

SYNONYM *Sedum* 'Sunset Cloud'

Really vivid—the flower petals open pink but are quickly eclipsed by large and showy carpels that deepen to a rosy-red. When in bloom, these flowerheads are prolific enough to mostly cover the entire plant. The abundant foliage, meanwhile, is blue-green and rounded, about 1 inch (2.5 cm) wide and a little longer. Stems are maroon red and initially upright, but then arching and sprawling. Deciduous. Propagate by division.

ZONES 4–9

PLANT SIZE 10–12+ inches (25–30 cm) tall, 12–14+ inches (30–36+ cm) across.

SOIL dry to well-drained, average

LIGHT full sun

SIMILAR SPECIES AND CULTIVARS Quite similar to *Hylotelephium* 'Bertram Anderson', though it grows taller and is, unfortunately, more susceptible to fungal disease. *Hylotelephium* 'Amber' has similar coloring but is bushier; 'Dazzleberry' has a fuller sprawling habit; 'Ruby Glow' has similar coloration but is shorter.

ORIGIN garden origin, bred by Jim Archibald in England.

LANDSCAPE AND DESIGN USES This plant's medium size makes it perfect for combining with both smaller and taller plants. It is excellent with fall asters.

Hylotelephium tatarinowii

SYNONYM *Sedum tatarinowii*

A small but appealing Asian species. It is easily recognized by its silver to gray, small, strongly toothed foliage. In summer, the leaves may gain purple edges, and fall color is yellow. This foliage is relatively thin, typically only ¼ inch (6 mm) wide and 1 inch (2.5 cm) long. The strongly clumping plants have arching stems; they tend to have lost some lower foliage by the time the plants bloom in late summer. The pretty white flowers with pronounced maroon anthers are held on thin pedicels, and hints of pink develop as they age. Seedheads age from green to cinnamon brown. Also, this particular sedum has substantial roots, especially for its size. Deciduous. Propagate by seed or division.

ZONES 4–9
PLANT SIZE 5–6 inches (13–15 cm) tall, 8 inches (20 cm) across.
SOIL rocky, dry to well-drained

LIGHT full sun

SIMILAR SPECIES AND CULTIVARS *Hylotelephium tatarinowii* 'Mongolian Snowflakes' seems to be more substantial than the species and has the added benefit of uniformity, which comes from being vegetatively propagated. The following hybrids are approximately comparable: *H.* 'Pure Joy', *H.* 'Soft Cloud', and *H.* 'Thundercloud'.

ORIGIN North China and Central Mongolia (Harlan Hamernik brought back plants from Mongolia in the late 1990s); Bluebird Nursery in Nebraska introduced both the species and the vegetative form to the trade.

LANDSCAPE AND DESIGN USES Due to its small size, this plant would not be ideal in a border. A small raised bed, rock garden, or a scree or gravel garden is best. On the edge of a wall, it looks nice with *Campanula poscharskyana* and *Saponaria ocymoides*. It also does well as a specimen in a container, with or without similar-size companions.

Hylotelephium telephium 'Black Beauty'

Black Beauty stonecrop
SYNONYM *Sedum telephium* 'Black Beauty'

Compared to many of the purple-leaved forms, this one is on the compact side and a definite improvement. New growth is a blend of purple-pink and blue-gray, with a glossy to satin texture. As the foliage ages, it turns green but by fall darkens considerably to purple-red. The stems, meanwhile, are maroon-red. Pink to red buds open in early autumn to reveal wide-cupped petals of creamy white suffused with pink or red, with wine-red carpels (extra-short stamens are tucked in next to the ovary). Flowerheads are to scale, measuring 2–3 inches (5–7.5 cm) wide, allowing foliage to peak through. Deciduous. Patented; propagation is prohibited.

ZONES 3–9
PLANT SIZE 18 inches (45 cm) and at least that much across.
SOIL dry to well-drained, average
LIGHT full sun to part shade
SIMILAR SPECIES AND CULTIVARS *Hylotelephium telephium* 'Desert Red' and 'Picolette'.
ORIGIN bred by Florensis in the Netherlands.
LANDSCAPE AND DESIGN USES In the fall border, these plants combine easily with asters and ornamental grasses.

Hylotelephium telephium 'Cherry Truffle'

SYNONYM *Sedum telephium* 'Cherry Truffle'

Improved color and a more compact growth habit make this relatively new introduction very enticing. Foliage is thick, dark red, and slightly notched; initially brown-green, it quickly turns red. Medium-size deep pink to ruby flowerheads are to scale with the plant size and the foliage. The plant has an airy profile, which helps keep the foliage clean and healthy. Deciduous. Patented; propagation is prohibited.

ZONES 4–9
PLANT SIZE 18 inches (45 cm) tall, 24 inches (60 cm) across.
SOIL well-drained is best
LIGHT full sun
SIMILAR SPECIES AND CULTIVARS *Hylotelephium telephium* 'Black Beauty'; 'Raspberry Truffle' is similar but shorter.
ORIGIN Terra Nova Nurseries, introduced in 2013.
LANDSCAPE AND DESIGN USES Due to the dark foliage, you will want to plant this with some lighter colors. I recommend grasses like little bluestem (*Schyzachrium scoparium*).

Hylotelephium telephium subsp. maximum 'Sunkissed'

SYNONYM *Sedum telephium* subsp. *maximum* 'Sunkissed'

Olive green foliage and white flowerheads make this one special. The leaves are spoon-shaped and have a semiglossy texture to them. The stems are lighter green. In late summer, light green buds open to white blooms accented with creamy yellow carpels. Deciduous. Patented; propagation is prohibited.

ZONES 3–9

PLANT SIZE 24 inches (60 cm) tall and at least that much across.

SOIL dry to well-drained

LIGHT full sun

SIMILAR SPECIES AND CULTIVARS *Hylotelephium telephium* subsp. *maximum* 'Bronco' is creamy white with a rose pink blush, and 'Gooseberry Fool' is among the whitest autumn stonecrops (although the latter can have red stems). *Hylotelephium telephium* 'Desert Blonde', new in 2013 from Terra Nova Nurseries, is very compact, only growing to 9 inches (23 cm) tall. New to the trade is *H. pallescens*, a white-flowered species from Russia with smaller green foliage and upright stature.

ORIGIN bred by Herbert Oudshoorn of Future Plants.

LANDSCAPE AND DESIGN USES I always like the effect that the creamy white autumn stonecrops have next to the red and pink forms, as if you're watching the evolution of the plants from one to another. This variety would also work nice in a more formal border where white, silver, or neutral colors are a focus and the form and color can get repeated. Or try it with some bright late-summer colors such as short yellow daylilies (*Hemerocallis*) or orange butterfly weed (*Asclepias tuberosa*).

Hylotelephium telephium 'Moonlight Serenade'

SYNONYM *Sedum telephium* 'Moonlight Serenade'

Gorgeous—the light-colored flowers make a statement against the darker background of the smoky purple foliage. A more compact habit is another distinct feature. The foliage emerges green, later deepening to more gray with some darker highlights. Maroon stems hold mauve-pink buds. The flowers open creamy white with creamy yellow carpels that eventually darken to rose purple for a riveting two-tone look. Deciduous. Patented; propagation is prohibited.

ZONES 3–9
PLANT SIZE 14–18 inches (46–45 cm) tall, 20+ inches (50+ cm) across.
SOIL dry to well-drained, average
LIGHT full sun
SIMILAR SPECIES AND CULTIVARS Other *Hylotelephium telephium* choices with bicolor flowers include 'Rainbow Xenox' (open creamy white but age mauve, above purple foliage) and 'Strawberries and Cream' (red and white, with dark green foliage). *Hylotelephium telephium* 'Stewed Rhubarb Mountain' is also comparable.
ORIGIN Future Plants in the Netherlands, introduced around 2010.
LANDSCAPE AND DESIGN USES Try it with another durable purple beauty, purple ironweed (*Vernonia lettermanii*). It would also look great alongside a fall grass like dwarf fountain grass (*Pennisetum alopecuroides* 'Piglet').

Hylotelephium telephium 'Picolette'

SYNONYM *Sedum* 'Picolette'

You get quite a color show with this beauty. Chocolate red foliage sports brown highlights; by the time the flowers emerge in late summer, it mellows to deep olive green. The maroon red stems are on the short side for a border sedum. Gray buds open on smaller, round flower clusters, 2–3 inches (5–7.5 cm) wide in opposite pairs. Creamy white petals age to light pink and surround rose pink carpels, which give these flowers most of their color. The brown seed heads have good substance and hold on into winter. Deciduous. Patented; propagation is prohibited.

ZONES 4–9
PLANT SIZE 12–15 inches (30–38 cm) tall, 18+ inches (45+ cm) across.
SOIL dry to well-drained
LIGHT full sun
SIMILAR SPECIES AND CULTIVARS *Hylotelephium telephium* 'Cherry Truffle' and 'Desert Red'
ORIGIN garden, introduced by Future Plants in the Netherlands.
LANDSCAPE AND DESIGN USES Due to its intermediate size, this form is appropriate with some of the groundcover sedums. Pairing it with a similarly colored form around the base, like *Sedum album* 'Coral Carpet', would make a stunning combination—it would almost look like the plant had self-seeded. Alternatively, short grasses such as blue fescue (*Festuca ovina*) are an easy complement to the dark foliage.

Hylotelephium telephium 'Postman's Pride'

SYNONYM *Sedum telephium* 'Postman's Pride'

The darkness of this plant's leaves are almost unmatched. Chocolate red, with a glossy to satin finish, they're wavy and tend to curl back at the tips. Dark buds match the color of the foliage—another unusual, and desirable, feature. The flowers have some creamy white to them in the beginning, but end up more on the pink side with reddish carpels. As they fade, the flowerheads darken to brick red. Deciduous. Patented; propagation is prohibited. This plant won an award at Planetarium, a European Horticultural Trade Show, in 2005.

ZONES 3–9
PLANT SIZE 24 inches (60 cm) tall and at least that much across.
SOIL dry to well-drained, average
LIGHT full sun
SIMILAR SPECIES AND CULTIVARS *Hylotelephium telephium* 'Jose Aubergine' is a sister seedling that was released in 2006. It does seem to be a better garden plant, but perhaps due to the name is not as widely grown yet. Other similar ones include *Hylotelephium telephium* 'Desert Black', 'Cherry Truffle', and 'Purple Emperor'.
ORIGIN garden origin; it was raised in 1999 by Jose Buck, a Belgian postman, as a controlled cross between *Hylotelephium telephium* 'Lynda et Rodney' and 'Purple Emperor'.
LANDSCAPE AND DESIGN USES This selection benefits from pinching early in the season to encourage branching and more compact growth. Lean soil will also help keep a plant from opening up or flopping. The upright habit matches nicely in autumn with autumn moor grass (*Molinia autumnalis*).

Hylotelephium telephium 'Purple Emperor'

SYNONYM *Sedum telephium* 'Purple Emperor', *S.* 'Washfield Purple', *S.* 'Washfield Ruby'

This older variety is one of the most widely available purple- to red-leaved and pink-purple flower types; it still a good plant choice compared to many of the red-leaved varieties that came after it. Upright and somewhat open growth habit is characteristic. New growth may emerge gray, but matures to a burgundy purple with a satiny or semiglossy finish. Leaves are mostly opposite, typically measuring over 1 inch (2.5 cm) wide and more than 2 inches (5 cm) long. Flowerheads are not large, reaching 2 inches (5 cm) across, but are usually in clusters of 4 or more, which increases their impact. Both the buds and flower color vary depending on site, exposure and temperatures. The pink buds can open with nearly white flowers, but bicolor flowers of rose are most common; with cool weather, they quickly deepen to wine red. Stamens are as long as petals. Deciduous. Propagate by cuttings or spring division.

ZONES 3–9

PLANT SIZE 18–24 inches (45–60 cm) tall, 24+ inches (60+ cm) across.

SOIL dry to well-drained, average

LIGHT full sun to part shade

SIMILAR SPECIES AND CULTIVARS Both *Hylotelephium telephium* 'Black Beauty and 'Desert Black' are more compact. *Hylotelephium telephium* 'Cherry Truffle' and 'Raspberry Truffle' are similar, but with improved form and habit.

ORIGIN garden origin; introduced by the now-defunct Washfield Nursery in England.

LANDSCAPE AND DESIGN USES All the *Hylotelephium telephium* cultivars require good drainage to perform their best, and this is no exception. The purple foliage looks exceptional next to the silver foliage of *Artemisia*, or yellow flowers.

Hylotelephium telephium 'Raspberry Truffle'

SYNONYM *Sedum telephium* 'Raspberry Truffle'

A real stunner—robust, colorful, and compact. The show begins in spring, when the plant's leaves emerge mid to dark green with red margins, soon turning entirely purple-red; they are opposite and nicely toothed. Buds and flowers are both raspberry pink with hints of cream in the background and they even have a fiery touch of orange-yellow in the carpels. Flowerheads start to open in late summer, and are made up of several smaller clusters held close together instead of one big dome. Deciduous. Patented; propagation is prohibited.

ZONES 4–10

PLANT SIZE 10+ inches (25+ cm) tall, 12+ inches (30+ cm) across.

SOIL dry to well-drained, average (I lost one of my two plants to disease and wet soil, but with good drainage, there should not be an issue)

LIGHT full sun

SIMILAR SPECIES AND CULTIVARS *Hylotelephium telephium* 'Black Beauty' is taller but has similar foliage color; 'Cherry Truffle' is more compact but similar; 'Desert Black' is shorter but similar color combination.

ORIGIN Terra Nova Nurseries, introduced around 2011.

LANDSCAPE AND DESIGN USES The combination of red foliage and raspberry-colored flowers is stunning all by itself, but this cultivar also mixes easily with many favorite border plants. Try it with catmint for its silver foliage.

Hylotelephium telephium 'Red Cauli'▾

SYNONYM *Sedum telephium* 'Red Cauli'

A newer variety that I predict is bound for great popularity, as its color is magnificent. Arching purple-red stems contrast with lighter gray-green, oblong foliage to make a unique statement in the garden. Buds are green, and open in late summer to deep raspberry red, with all parts having the same hue (but no stamens). Each inflorescence is around 4 inches (10 cm) wide. Deciduous. Propagate by division or cuttings. Received the Award of Garden Merit (AGM) from Britain's Royal Horticultural Society in 2006.

ZONES 4–9
PLANT SIZE 24–28 inches (60–70 cm) tall, 36+ inches (90+ cm) across.
SOIL dry to well-drained, average
LIGHT full sun
SIMILAR SPECIES AND CULTIVARS *Hylotelephium telephium* 'Desert Red' is shorter, and 'Marchants Best Red' has similar arching stems but redder foliage. One of the parents was *H. telephium* subsp. *telephium* 'Munstead Dark Red'; I suspect *H. telephium* subsp. *ruprechtii* as the pollen parent and source of its arching habit.
ORIGIN introduced by Graham Gough of Marchants Hardy Plants in the United Kingdom.
LANDSCAPE AND DESIGN USES A low planting of purple lovegrass, *Eragrostis spectabilis*, around these plants would be a stunning sight.

Hylotelephium telephium subsp. *ruprechtii* 'Hab Gray'▾

SYNONYM *Sedum telephium* subsp. *ruprechtii* 'Hab Gray'

Though its name promises gray, the distinctive glaucous leaves and upright, arching stems of this selection vary from silver to blue, depending on the soil and exposure. The leaves have toothed edges, appear opposite, and clasp the stems. Flower clusters, carried at the tips of the stems, are on the small side, about 2 inches (5 cm) across. Petals are creamy while, ovaries are yellowish, and the flowers fade to an overall creamy white. Deciduous. Propagate by division.

ZONES 3–9
PLANT SIZE 20–28 inches (50– 70 cm) tall, 20+ inches (50+ cm) across.
SOIL dry to well-drained, average
LIGHT full sun to light shade
SIMILAR SPECIES AND CULTIVARS *Hylotelephium telephium* 'Desert Blonde' is more compact but otherwise similar. *Sedum caucasicum* is a short blue-green species that also has gray foliage and white flowers; unfortunately, it is rare in cultivation. 'Hab Gray' is probably similar to the species, which is not commonly sold.
ORIGIN garden origin
LANDSCAPE AND DESIGN USES This is a great plant in the right location due to its tall and upright stature. I recommend growing it in lean soil or in an extra-well-drained to dry site, which keeps it shorter (plus it will be less prone to rotting if there are wet periods). An interesting duo would be pairing this bluish plant with one of the blue-leaved cultivars of little bluestem (*Schizachyrium scoparium*).

Hylotelephium telephium subsp. *telephium* 'Munstead Dark Red'▲

SYNONYM *Sedum telephium* subsp. *telephium* 'Munstead Dark Red'

An interesting and good-looking plant. Its gray-green foliage typically has some purple to red highlights on the edges and by the time it blooms, the stems usually but not always deepen in hue. The flowers are a bit curious: petals begin pink, enrich to red, and are accompanied by darker, extra-large carpels that continue to darken after the petals have faded. Stamens are either absent or quite short, which suggests that the plant may be sterile and possibly a hybrid. The red seedheads look good well into the winter months. Deciduous. Propagate by division.

ZONES 3–9

PLANT SIZE 18 inches (45 cm) tall, 24+ inches (60+ cm) across.

SOIL well-drained, average

LIGHT full sun

SIMILAR SPECIES AND CULTIVARS *Hylotelephium telephium* 'Stewed Rhubarb Mountain' is a newer, improved form from England. *Hylotelephium telephium* 'Abbey Dore' and 'Herbstfreude' are comparable. *Hylotelephium telephium* 'Desert Red' is shorter.

ORIGIN named by Gertrude Jekyll in England.

LANDSCAPE AND DESIGN USES This dependable plant can be used in a mixed perennial border for late summer color as well as winter interest. Combine it with ornamental grasses and other plants that peak in fall, like asters.

Hylotelephium telephium 'Xenox'

SYNONYM *Sedum telephium* 'Xenox', *Hylotelephium telephium* 'Karlfunkelstein'

Some consider this an improved form of *Hylotelephium telephium* 'Purple Emperor', and it is indeed larger and more vigorous. Olive green foliage early on turns plum (with a hint of green in the background). The leaves are substantial, lightly toothed, and almost clasping on the maroon red stems. The petals of the flowers are noticeably short and wide, mostly creamy white with some pink; they contrast strongly with the raspberry-rose colored carpels. Deciduous. Patented; propagation is prohibited.

ZONES 3–9
PLANT SIZE 18 inches (45 cm) tall and at least that much across.
SOIL dry to well-drained, average
LIGHT full sun
SIMILAR SPECIES AND CULTIVARS Most like *Hylotelephium telephium* 'Purple Emperor' or 'Raspberry Truffle'; 'Black Beauty' is more compact. *Hylotelephium telephium* 'Rainbow Xenox' is a newer form with a blend of creamy white flowers and some red carpels emerging from reddish buds on red stems (its habit can be open, however, so plants should be grown on dry, lean soil for best performance). *Hylotelephium* 'Orange Xenox' has dark green foliage, red stems, and flowers open to cream petals suffused with orange red.

ORIGIN garden origin; bred in 2002 by Hubert Oudshorn of the Netherlands.

LANDSCAPE AND DESIGN USES This plant is perfect for fall borders. It's excellent among ornamental grasses like *Calamagrostis* × *acutiflora* 'Karl Foerster' or *Panicum virgatum* 'Northwind', and vivid purple perennials such as *Symphyotrichum* (Aster) *novae-angliae* 'Purple Dome'.

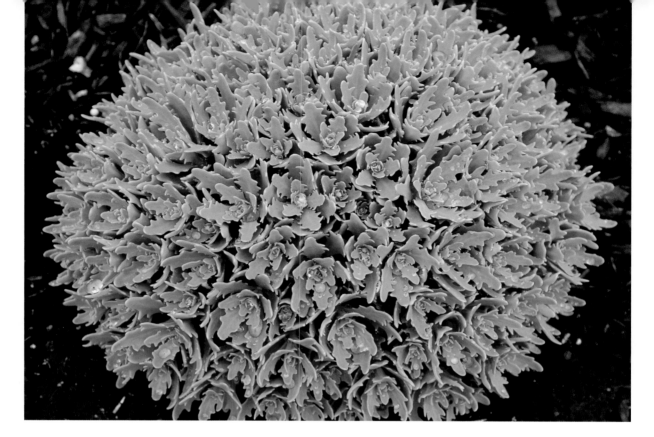

Hylotelephium 'Thundercloud'

SYNONYM *Sedum* 'Thundercloud'

This hybrid represents a fairly new look in sedum in habit, foliage, and texture. It forms full, dome-shaped mounds of extremely toothed leaves; they are generally silver-green, but gain a blue hue in rich soil. It billows into bud and flower in mid to late summer. Depending on the site and exposure, flowers are white to light pink; they usually have slight marks of pink either at the tips of the buds or at the tips of the white carpels. Petals are ridged, and anthers are dark maroon red. Deciduous. Patented; propagation is prohibited.

ZONES 4–9
PLANT SIZE 12 inches (30 cm) tall, 12–15+ inches (30–38+ cm) across.
SOIL dry to well-drained, average
LIGHT full sun

SIMILAR SPECIES AND CULTIVARS *Hylotelephium tatarinowii* looks similar but is smaller. *Hylotelephium* 'Pure Joy' makes one big flower dome of pink bicolor flowers.
ORIGIN developed by the author, introduced in 2009.
LANDSCAPE AND DESIGN USES The ball-shaped mounds have such a uniform appearance that they can be used in formal or very structured designs. This plant combines nicely with short grasses like *Sesleria autumnalis* or *S.* 'Greenlee's Hybrid', which bloom around the same time with creamy white, brushlike plumes. At the nursery, I have it underplanted with *Sedum album* var. *micranthum* 'Chloroticum' in a long row along a paver walkway. Across the path is another nifty combination; there, 'Thundercloud' joins long-blooming *Calamintha nepeta* subsp. *nepeta*'s light blue, practically white flowers—the soft color complements its white and pale pink hues.

Hylotelephium 'Thunderhead' ▲

SYNONYM *Sedum* 'Thunderhead'

Gray-green substantial foliage deepens to darker green as the stems increase. By late summer, the stems turn dark red and present extra-large, light green buds that open to wine purple flowers. Flowerheads are large, 6 inches (15 cm) across. Deciduous. Patented; propagation is prohibited.

ZONES 4–9
PLANT SIZE 30 inches (76 cm) tall, 24 inches (60 cm) across.
SOIL average, well-drained
LIGHT full sun
SIMILAR SPECIES AND CULTIVARS *Hylotelephium* 'Class Act', 'Dynomite', and 'Mr. Goodbud'.
ORIGIN Terra Nova Nurseries, introduced in 2012.
LANDSCAPE AND DESIGN USES This plant fits right into a fall border with blue and purple asters, Russian sage, the yellow fall foliage of amsonia, and ornamental grasses. The deep flower color also makes a fine complement to plum-leaved *Heuchera* 'Obsidian'.

Hylotelephium ussuriense 'Turkish Delight'

Turkish Delight stonecrop
SYNONYM *Sedum ussuriense* 'Turkish Delight'

A picture in regal shades of red. The flowers are a rich, saturated ruby and are carried in congested clusters for maximum attention. The medium-size plant has gray foliage that rapidly deepens to burgundy red; leaves are mostly round and about 1 inch (2.5 cm) across. Older foliage develops purple edges and becomes mostly dark by the time the bright flowers open in late summer to early autumn. Chalky or glaucous buds and petals surround purple-red carpels. Even the stems contribute to the show, also turning burgundy red by fall. Deciduous. Propagate by seed or division.

ZONES 3–10
PLANT SIZE 12–15 inches (30–40 cm) tall and at least that much across.

SOIL dry to rocky, well-drained, average
LIGHT full sun
SIMILAR SPECIES AND CULTIVARS *Hylotelephium* 'Amber' is quite possibly a hybrid with this plant, as it has a strong resemblance in terms of leaf shape and habit.
ORIGIN Eastern Siberia and Korea. In 1994, when Ray Stephenson wrote his book on sedums, this plant was not even in cultivation and it is currently endangered in Korea. Thompson & Morgan started selling seed around 2000, and since then it has trickled into the trade.
LANDSCAPE AND DESIGN USES The size of this plant makes it easy to combine with both the groundcover sedums and the taller, upright hylotelephiums. It is perfect for use with blue fescue.

Hylotelephium 'Vera Jameson'

SYNONYM *Sedum* 'Vera Jameson'

A lovely, medium-size plant in a red-purple theme. The gray-green foliage takes on a purple cast—a very distinctive feature of this variety. Stems are almost always red-purple. Buds are gray with a hint of purple, but open to pretty dark pink flowers.

ZONES 4–9
PLANT SIZE 8–12 inches (20–30 cm) tall, 10–12+ inches (25–30+ cm) across.
SOIL dry to well-drained, average
LIGHT full sun
SIMILAR SPECIES AND CULTIVARS *Hylotelephium* hybrids such as 'Dazzleberry', 'Plum Perfection,

'Ruby Glow', and 'Sunset Cloud'
ORIGIN made in the 1970s; probably a hybrid of *Hylotelephium telephium* subsp. *maximum* 'Atropurpureum' and *H.* 'Ruby Glow'.
LANDSCAPE AND DESIGN USES Devote an area to a patchwork-style groundcovering display featuring this one and other, complementary medium-size relatives such as *Hylotelephium* 'Bertram Anderson', 'Cherry Tart', 'Dazzleberry', or 'Ruby Glow'. Tuck in a few plants of the pink-flowered *Dianthus* 'Feuerhexe' (Firewitch) so there will be a splash of hot color earlier in the summer.

Orostachys boehmeri

SYNONYM *Sedum boehmeri, Orostachys aggregatus*

Glossy green rosettes, 1–2 inches (2–5 cm) across, are composed of round-tipped leaves. Many thin stolons (aboveground stems) emerge from between the leaves—a key identifying feature of this species. Very late-blooming: spikes of thick white flowers appear in mid to late autumn. Evergreen. Propagate by cutting off and replanting the "pups" (little plantlets.)

ZONES (5)6–9
PLANT SIZE 2–3 inches (5–7.5 cm) tall, 24 inches (60 cm) across, flower stems 6–8 inches (15–20 cm) tall.
SOIL dry, with excellent drainage
LIGHT full sun
SIMILAR SPECIES AND CULTIVARS If your plant has gray-green foliage, it is more likely the cultivar 'Keiko', which is often incorrectly sold under the species name. The similar *Orostachys furusei* has glaucous foliage on larger and more prolific plants. Its greenish white flowers contain many bracts.
ORIGIN Hokkaido and Honshu Islands, Japan.
LANDSCAPE AND DESIGN USES These do well planted in troughs or other containers, but they also work in a rock garden. They make stunning displays with other dunce caps, other *Orostachys* species, and hens and chicks, *Sempervivum* cultivars.

Petrosedum forsterianum subsp. *elegans*

SYNONYM *Sedum forsterianum* subsp. *elegans*

Texture makes this one stand out. Its extra-thin, gray-green (almost blue) glaucous foliage is soft to the touch. Before the yellow flowers open, the buds are reflexed and the foliage on the lower part of the stems starts to dry out. Evergreen. Propagate by division, cuttings, or seed.

ZONES (5) 6–9
PLANT SIZE 10 inches (25 cm) tall, 12+ inches (30+ cm) across.
SOIL moist to rocky
LIGHT light to part shade
SIMILAR SPECIES AND CULTIVARS *Petrosedum forsterianum* 'Blue Lagoon' and 'Oracle' are similar. *Petrosedum forsterianum* subsp. *elegans* 'Silver Stone' is a seed variety that has purple stems at the base, and forms more rounded rosettes (it may or may not be valid as different from the species). *Petrosedum forsterianum* subsp. *elegans* f. *purpureum* is from North Africa; its glaucous foliage has attractive purple highlights.
ORIGIN England, France, Belgium, (West) Germany, Spain, Portugal, and Morocco.
LANDSCAPE AND DESIGN USES Its finely textured, purple-hued foliage is synergistic and easy to combine with other plants such as *Campanula poscharskyana* or gentians (*Gentiana*). In Zone 6 or warmer, this can be a rapid-spreading groundcover. Not reliably hardy in Zone 5.

Petrosedum montanum subsp. *orientale*

SYNONYM *Sedum montanum* subsp. *orientale*

Forms a more substantial clumping mound, and is considered an improved form of *Petrosedum rupestre* 'Blue Spruce'. It tolerates heat and humidity better. Thick, strongly pointed, gray-green foliage is congested at the end of the stems, while the rest of the stem is bare. In winter, the leaves gain a purplish cast and appear more glaucous. It can be hard to tell this species apart from many of its relatives, but one way is the always-upright development of the flower bud, whereas *Petrosedum rupestre* will have reflexed or pendant buds. Individual flowers have six petals and are yellow. Evergreen. Propagate by cuttings.

ZONES 4–9
PLANT SIZE 8 inches (20 cm) tall, 12 inches (30 cm) across.
SOIL rocky, dry to well-drained, average

LIGHT full sun
SIMILAR SPECIES AND CULTIVARS *Sedum lanceolatum* is usually finer, smaller, and lighter gray-green. *Petrosedum montanum* (the species, not its cultivars) is also similar, though foliage color can vary from green to gray. *Petrosedum ochroleucum* is a slightly smaller plant but has the same upright flower buds. *Petrosedum rupestre* 'Blue Spruce' and *P. sediforme* are both comparable, but have larger leaves.
ORIGIN France, Switzerland, Italy, and the former Yugoslavia.
LANDSCAPE AND DESIGN USES This plant is ideal for use in a greenroof planting project. It also makes a great groundcover in the garden. Because it is evergreen and spreading, it is also often used with evergreen shrubs.

Petrosedum ochroleucum

SYNONYM *Sedum ochroleucum, S. anopetalum*

This species is almost identical to *Petrosedum rupestre* and can only be identified with complete certainty when in bud or flower stage; its buds and petals are erect. The forms I have grown have been less vigorous, less hardy, and shorter than *P. rupestre*. The leaves are also not as congested at the tips. Evergreen. Propagate by cuttings.

ZONES 5–10

PLANT SIZE 2–3 inches (5–7.5 cm) tall, 6+ inches (15+ cm) across, flower stems 4 inches (10 cm) tall.

SOIL rocky, dry to average

LIGHT full sun

SIMILAR SPECIES AND CULTIVARS *Petrosedum rupestre* is almost identical. *Petrosedum ochroleucum* 'Centaureus' is green with red highlights, while 'Red Wiggle' (see photo) has dark green foliage that turns mostly red in winter.

ORIGIN Spain, France, Italy, the former Yugoslavia, Albania, Greece, Bulgaria, and Turkey.

LANDSCAPE AND DESIGN USES I like the smaller stature and strong red highlights on the forms I have grown, but they are not reliably hardy or vigorous enough for me in Zone 5. It should be dependable in a rocky or well-drained site in Zone 6. Due to its native habitat, it should be tolerant of hot summers. In winter, the reddened foliage would be a nice contrast to *Petrosedum rupestre* 'Angelina'.

Petrosedum rupestre

Jenny's stonecrop,
crooked stonecrop
SYNONYM *Petrosedum reflexum, Sedum
reflexum, S. rupestre*

A handsome plant in all seasons! In winter, its
whorled, needlelike foliage has red tips and highlights. It greens up in summer. The vigorous habit is
creeping and sprawling, and the stems are reddish
brown. Upright flower stems can reach up to 12
inches (30 cm) tall. The buds are reflexed and open
to bright yellow flowers with six to seven petals, in
open sprays around 3 inches (7.5 cm) wide. Easy to
propagate from cuttings in spring and fall.

ZONES 4–9
PLANT SIZE 4–6 inches (10–15 cm) tall, 18–24
inches (45–60 cm) across, flower stems to 12
inches (30 cm) tall.
SOIL dry to well-drained, average
LIGHT full sun to part shade
SIMILAR SPECIES AND CULTIVARS *Petrosedum rupestre* 'Angelina' has gold foliage, while 'Blue Spruce'
has blue foliage. *Petrosedum rupestre* 'Green Spruce'
is lighter green and has more vigorous growth.
Petrosedum rupestre f. *cristatum* is a crested form
that unfortunately reverts easily; remove any non-
crested sports as they appear. *Sedum* ×*luteolum* is
a hybrid between *Petrosedum rupestre* and *S. sediforme*. *Sedum* ×*lorenzoi* is a hybrid between *Petrosedum rupestre* and *P. ochroleucum*. *Petrosedum rupestre* subsp. *viride* is a smaller version with a congested
habit and shorter flower stems; it bears lighter yellow flowers on chalky green foliage.
ORIGIN Central and Western Europe.
LANDSCAPE AND DESIGN USES Try it with other colors of sedum that also have needled foliage; such
a display makes for attractive contrasts. This one
also makes a good partner with prickly pear cactus
(*Opuntia humifusa*), because both are green and
have the same texture.

Petrosedum rupestre 'Angelina'

Angelina stonecrop
SYNONYM *Sedum rupestre* 'Angelina'

This sedum has exploded in popularity since it was introduced to the United States in 2000, and for good reason. It makes a great container plant contrasting nicely with most everything. It has good vigor, which can be an issue with some yellow-leaved plants. Its golden yellow foliage turns copper-orange in winter, making it easy to recognize. The overall habit is spreading and sprawling. Foliage is linear and needlelike, whorled around the stems. The plant can be shy to bloom, but if it does, yellow clusters appear in midsummer atop 8-inch (20-cm) stems. Evergreen. Easy to propagate from cuttings in spring and fall.

ZONES 4–9
PLANT SIZE 4–6 inches (10–15 cm) tall, 18–24 inches (45–60 cm) across, flower stems 8 inches (20 cm) tall.
SOIL rocky, dry to well-drained, average
LIGHT full sun

SIMILAR SPECIES AND CULTIVARS *Petrosedum rupestre* 'Sea Gold' and *Sedum mexicanum* 'Lemon Ball' (synonym 'Lemon Coral'), although they are not cold-hardy. There is an unnamed crested form, too.

ORIGIN found in a private garden in Croatia by Cristian Kress of Sarastro Nursery; he asked for a cutting. In 2000, Robert Herman of Connecticut brought it to the United States.

LANDSCAPE AND DESIGN USES This plant can be used as an effective edging plant between a paved surface and a garden bed. It is also commonly used on greenroofs. One of the easiest plants to combine with, it looks terrific with short Shasta daisies like Daisy May *Leucanthemum superbum* 'Daisy Duke'. Also combines well with red-leaved plants, including ajugas, heucheras, and sempervivums. Its gold foliage complements yellow-flowered plants with ease, too, for instance, *Coreopsis verticillata* 'Moonbeam'.

Petrosedum rupestre 'Blue Spruce'

Blue Spruce stonecrop
SYNONYM *Sedum rupestre* 'Blue Spruce'

This popular, adaptable plant features blue needlelike foliage on vigorous spreading plants that remind me of an evergreen. In winter, some of the old foliage can take on a purple color below the tips. Buds are reflexed and straighten up just before they open to reveal a bright yellow flower with six to seven petals. These petals are rounded at their tips (like the flowers of similar *Sedum sediforme*). The sprawling, red-brown stems tend to be bare at the base. Evergreen. Easy to propagate from cuttings in spring and fall.

ZONES 4–9
PLANT SIZE 4–6 inches (10–15 cm) tall, 18–24 inches (45–60 cm) across.
SOIL dry to well-drained, average (I've seen this plant naturalized on Stone Mountain in Georgia, growing on a stone face with barely any soil.)
LIGHT full sun to part shade
SIMILAR SPECIES AND CULTIVARS *Petrosedum montanum* subsp. *orientale* has smaller leaves. *Petrosedum sediforme* has larger foliage and light yellow flowers. *Petrosedum rupestre* 'Nanum' is a choice smaller blue form with more compact growth. It was pointed out to me by Scott Ogden that the flowers of this plant seem to be intermediate between *P. rupestre* and *P. sediforme*, so it could be a hybrid.
ORIGIN garden origin
LANDSCAPE AND DESIGN USES This is a very common form for good reason; the blue-green color and ease of culture make it a versatile plant. It is used on greenroofs, as a groundcover, in mixed containers, in hanging baskets, and in rock gardens. The extreme heat of summer slows down its growth, so in hotter climates, be sure to provide some shade.

Petrosedum sediforme

SYNONYM *Sedum sediforme, Petrosedum nicaeense*

A more substantial, taller plant than any of the other needlelike forms. Also, unlike its relative *Petrosedum rupestre*, its foliage is flat on top. The thick linear leaves vary in color from a range of greens to blue-green, and the stems can be as thick as ⅛ inch (3 mm) or more. Flower stalks are proportionally taller as well and carry round clusters at the tops in soft creamy yellow. Evergreen. Propagate by cuttings.

ZONES 5–10
PLANT SIZE 8–10 inches (20–25 cm) tall, 24 inches (60 cm) across, flower stalks to 20 inches (50 cm).
SOIL dry to well-drained, average
LIGHT full sun
SIMILAR SPECIES AND CULTIVARS Strongly resembles *Petrosedum rupestre*, which has gold flowers and a smaller stature. Similar relatives include *P. montanum* and

P. ×*thartii* (a hybrid between *P. sediforme* and *P. montanum*). Vegetative forms are beginning to be sold in the United States. 'Turquoise Tails' is a large, chalky blue selection from Colorado (in general, the blue ones look like smaller forms of burro's tail, *Sedum morganianum*).
ORIGIN Southern Europe, North Africa, even Turkey, Syria, and Israel—areas near the Mediterranean coast.
LANDSCAPE AND DESIGN USES The substantial foliage and variability of this species allow it to be used as a transition from small forms to longer-leaved sedums and other larger plants. Although it is not grown widely in North America yet, I believe it has great potential, especially in the South because it is extremely heat- and drought-tolerant. The more substantial size makes it an ideal companion in mixed containers.

Phedimus aizoon

SYNONYM *Sedum aizoon*

This species has some of the freshest looking foliage in the spring. Leaves are also comparatively large, reaching 2 inches (5 cm) long and over 1 inch (2.5 cm) wide. Texture is glossy, with leaf color varying from pale green to dark green. In spring, upright stems emerge and grow rapidly, reaching over 12 inches plus (30+ cm) tall by early summer, when they are topped by flattened yellow flower clusters 3–4 inches (7.5–10 cm) wide. Stem color is also variable, although the nicest forms are red-stemmed. The stems die back to the ground for the winter. For the following year, new buds overwinter as rosettes at ground level. Deciduous. Best increased by division in early spring; also, reseeds easily.

ZONES 4–9
PLANT SIZE 15–29 inches (38–73 cm) tall, 20–24 inches (50–60 cm) across.
SOIL moist, average
LIGHT full sun to part shade
SIMILAR SPECIES AND CULTIVARS There are a number of similar-looking relatives, including: *Phedimus aizoon* var. *latifolium*, which has broad leaves, *P. aizoon* subsp. *angustifolium*, a narrow-leaved form, and *P. aizoon* 'Aurantiacum' ('Euphorbioides'), a form with red stems and orange-yellow flowers. Closely related *P. maximowiczii* has darker foliage and smaller clasping leaves, and is easily confused with this one. *Phedimus litoralis* is an intermediate between *P. aizoon* and *P. kamtschaticus*, growing to 12 inches (30 cm) tall.
ORIGIN Siberia, to the east coast, Mongolia to Japan.
LANDSCAPE AND DESIGN USES Note that this plant has its peak early in the season, so it should be mixed with late-blooming plants to distract from its fading appearance. Because the foliage is so bright, it looks great with other colorful relatives, for instance, alongside the dark leaves and red stems of *Phedimus aizoon* 'Aurantiacum' (synonym 'Euphorbiodes'). It would also make a nice companion for low-growing *Amsonia* 'Blue Ice'. This plant may get overlooked because so many other yellow flowers bloom at the same time of year, but it's a good plant that deserves more use. Considering its size, it is very drought tolerant.

Phedimus ellacombeanus ▲

SYNONYM *Sedum kamtschaticum* var. *ellacombianum*

This strong grower has excellent drought tolerance, is cold hardy, and makes a fine-looking groundcover throughout the growing season. Its glossy leaves are toothed on the top third. In spring and summer, foliage is light, bright green, joined in summer by substantial yellow flower clusters around 3+ inches (7+ cm) wide. Fall foliage is superior—hints of orange, leaning to red, while the seedheads are distinctly orange, maturing to red. The plants overwinter as a tight cushion of green rosettes at ground level (the stems are deciduous). Increasing plants from division is easy.

ZONES 4–9, probably hardier
PLANT SIZE 8 inches (20 cm) tall, 12 inches (30 cm) across.
SOIL dry to well-drained, average to moist
LIGHT full sun to nearly full shade
SIMILAR SPECIES AND CULTIVARS Sometimes mislabeled and sold as *Phedimus selskianus*, a relative that is hairy in all its parts. Other similar plants include *P. kamtschaticus*, *P. hybridus* 'Immergrünchen', and *P. takesimensis* 'Golden Carpet'.
ORIGIN all European countries (except Ireland and Iceland), extending into North Africa.
LANDSCAPE AND DESIGN USES This plant makes an excellent groundcover. It's very adaptable, growing surprisingly well in part to nearly full shade.

Phedimus ellacombeanus 'The Edge' ▶

SYNONYM *Sedum kamtschaticum* var. *ellacombianum* 'The Edge', *S. kamtschaticum* 'The Edge'

A very beautiful variegated plant! It's at its peak in spring, when the yellow edge against the chartreuse green foliage is bold and bright. The contrast tends to fade as the season progresses, and the fall color is not as distinct as the species. Although it is not as vigorous as the all-green form, it is still a robust plant. Golden yellow flowers bloom in summer in small bunches. Strongly deciduous. As for propagation, cuttings are relatively easy, but division will work, too.

ZONES 4–9
PLANT SIZE 6–8 inches (15–20 cm) tall, 10 inches (25 cm) across.
SOIL dry to well-drained, average
LIGHT full sun to part shade
SIMILAR SPECIES AND CULTIVARS *Phedimus kamtschaticus* 'Variegatus' is somewhat similar. This plant is occasionally listed as *Sedum selskianum* 'Variegatum', but that is wrong on two counts: all parts of the plant are not hairy, and the name for that species is now correctly *Phedimus selskianus*. It comes by its name confusion honestly. I got my original plant from Ed Scrocki, who sent it as another form of variegated *P. kamtschaticus*, to which this species is closely related (in fact, this species used to be considered a subspecies of it). To distinguish it from the other form of variegated *P. kamtschaticus*, I attached the name of 'The Edge' and it stuck. I see it commonly used in the trade now.
ORIGIN garden origin
LANDSCAPE AND DESIGN USES This plant is extremely drought-tolerant. It makes a nice companion for other yellow-flowered plants of similar size, such as *Corydalis lutea* and *Euphorbia myrsinites*. As with any variegated sedum, reversion is a possibility; if you spot any all-green sports, cut them out as close to the crown of the plant as possible.

Phedimus hybridus 'Immergrünchen' ▲

SYNONYM *Sedum hybridum* 'Immergrünchen'

Here is a fast-growing, substantial plant that forms handsome, spreading clumps. Its leaves are comparatively wider than its close relatives (mentioned below); the fact that they are completely evergreen is another telltale sign for this species. Notches on the top half of the leaves also seem to be more pronounced. Over time, the lower portions of the stems become bare. Clusters of yellow flowers, appearing in early summer, are heavy but not numerous. Evergreen. Easy to propagate by cuttings or division.

ZONES 4–9
PLANT SIZE 6–8 inches (15–20 cm) tall, 15–20 inches (38–50 cm) across.
SOIL rocky, well-drained, average
LIGHT full sun
SIMILAR SPECIES AND CULTIVARS *Phedimus kamtschaticus*, *P. ellacombeanus*, and *P. takesimensis* 'Golden Carpet'
ORIGIN garden origin
LANDSCAPE AND DESIGN USES Thanks to its rapid growth rate, this is an ideal groundcover. (I received my first cutting of this plant during the yearly Sedum Society cutting exchange; within a year's time I had hundreds of plants!) Plant it with more substantial perennial plants like *Achillea* and *Geranium sanguineum*.

Phedimus kamtschaticus

Orange stonecrop,
Russian stonecrop
SYNONYM *Sedum kamtschaticum*

While this form is one of the most common, the best way to distinguish it is by comparing it to its relatives. The foliage is medium green in color and matte in texture compared to *Phedimus ellacombeanus*, which is lighter green and glossy. Foliage size is usually two to three times as long as wide. The top halves of the leaves are toothed in early summer. While the plant's habit is open and loose, you still get a strong clump. Each tip develops a bud that opens to a golden yellow, 2- to 3-inch (5- to 7.5-cm) diameter flower cluster. Fall color can be brief, but is stunning orange red—hence the common name. Deciduous. Easy to propagate from division, seed, and cuttings.

ZONES 4–9
PLANT SIZE 6–7 inches (15–18 cm) tall, 12 inches (30 cm) across.
SOIL stony, dry to well-drained, average
LIGHT full sun
SIMILAR SPECIES AND CULTIVARS *Phedimus hybridus* 'Immergrünchen' is fairly similar; *P. kamtschaticus* 'Variegatus' has foliage with creamy yellow edges and a slower growth rate.
ORIGIN Siberia to central China.
LANDSCAPE AND DESIGN USES This plant is very adaptable and widely used as both a groundcover and on greenroofs. It's also a fine garden citizen. I have it planted with bloody cranesbill, *Geranium sanguineum*, whose magenta-purple flowers appear at about the same time.

Phedimus kamtschaticus var. *floriferus* 'Weihenstephaner Gold'

Russian stonecrop

SYNONYM *Sedum kamtschaticum* var. *floriferum* 'Weihenstephaner Gold', *S. floriferus* 'Weihenstephaner Gold'

One of the best sedums in many regards! Slender, dark green foliage and strongly notched tips are the first clue in identifying this plant. From there, these floriferous plants have extra-branched red stems terminating in golden yellow flowers beginning in early summer. As the flowers fade, the seedheads develop some red tones, which give an appealing red and yellow bicolor effect. And finally, the beet-red foliage in winter sets this one apart from any imposters. Easy to propagate by cuttings.

ZONES 3–9

PLANT SIZE 5–8 inches (13–20 cm) tall, 5–6 inches (13–15 cm) across.

SOIL dry to well-drained, average; prefers summer moisture

LIGHT full sun to part shade

SIMILAR SPECIES AND CULTIVARS *Phedimus hybridus* 'Czar's Gold', which is grown from seed, does not have the same quality red winter coloration, but otherwise is quite similar. Leaves are slightly wider, more like *P. kamtschaticus*. *Phedimus middendorffianus* var. *diffusus* also resembles this form, but is not evergreen. *Phedimus sichotensis* also has similarities.

ORIGIN garden origin

LANDSCAPE AND DESIGN USES An outstanding plant, suitable as a groundcover, on greenroofs, and in containers. This is always my first recommendation if someone is looking for a low-growing groundcover. The true evergreen nature of this plant makes it foolproof. It will keep out weeds and the red winter color is great.

Phedimus middendorffianus

SYNONYM *Sedum middendorffianum*

Prized for its vivid red fall color, which can be some of the brightest imaginable. In spring and summer, the foliage is rich green. It's relatively thin at ¼ inch (6 mm) wide, with a toothed tip. In flower and form, this plant resembles *P. kamtschaticus* var. *floriferus* 'Weihenstephaner Gold', but it has a more relaxed habit and—if you look closely when it is in bloom—the carpels have extra-short beaks. Gold blooms open in summer. Deciduous. Propagate in spring by cuttings or division.

ZONES 4–9
PLANT SIZE 5–6 inches (13–15 cm) tall, 12 inches (30 cm) across.
SOIL dry to well-drained, stony or average
LIGHT full sun
SIMILAR SPECIES AND CULTIVARS Frequently sold as *Phedimus middendorffianus* var. *diffusus*, but this form is thinner with intermediate foliage, more like *P. middendorffianus* 'Striatus'. Others that are somewhat similar include *Phedimus hybridus* 'Czar's Gold', *P. kamtschaticus* var. *floriferus* 'Weihenstephaner Gold', and *P. sichotensis*.
ORIGIN Siberia, China, and Japan.
LANDSCAPE AND DESIGN USES The fall color on this plant is so spectacular, I would grow it for that reason alone. It also makes a great groundcover even though it's not evergreen (however, it does overwinter as green rosettes at ground level). Site it with plants that remain evergreen such as candytuft (*Iberis sempervirens*) or rock cress (*Arabis caucasica*).

Phedimus middendorffianus 'Striatus'

SYNONYM *Sedum middendorffianum* 'Striatum'

The fine texture and great fall color are my favorite attributes. Slender leaves of dark green always have some chocolate brown to red edges and highlights, which give this species a unique look. The golden yellow flowers are not reliable or very showy. Plants are strongly clumping. Deciduous. Propagate by division.

ZONES 4–9
PLANT SIZE 4 inches (10 cm) tall and wide.
SOIL stony and dry to well-drained, average
LIGHT full sun
SIMILAR SPECIES AND CULTIVARS *Phedimus hybridus* 'Czar's Gold', *P. middendorffianus* and *P. middendorffianus* var. *diffusus*, and *P. kamtschaticus* var. *floriferus* 'Weihenstephaner Gold'.
ORIGIN garden origin
LANDSCAPE AND DESIGN USES This plant is on the small side, so it will fit in a rock garden with other small clumping plants. Plants can be so dark at times that they tend to blend right in with the soil, so use where it can be complemented by light-colored gravel mulch.

Phedimus obtusifolius var. *listoniae*

SYNONYM *Sedum obtusifolium* var. *listoniae*

One of my favorite sedums! Its leaves are bright apple green. The plants overwinter as rosettes 3–4 inches (7.5–10 cm) wide, with the older foliage taking on some brownish red highlights at that time. In summer, short, branching flower stalks sport pretty flowers: deep purple-pink, with lightly cupped petals that switch abruptly to white at the base. Seedheads dry to cinnamon brown, and foliage can fade away in the heat of summer. In fall, with a little digging, you can observe unique dormant white buds, ready to sprout a new rosette. Evergreen. Propagate by division or seed.

ZONES (4)5–9

PLANT SIZE 3–4 inches (7.5–10 cm) tall, 8–10 inches (20–25 cm) across.

SOIL well-drained, average

LIGHT full sun

SIMILAR SPECIES AND CULTIVARS Annual *Phedimus stellatus* has foliage with a more ruffled edge.

ORIGIN Northwestern Anatolia (eastern Turkey).

LANDSCAPE AND DESIGN USES A perfectly hardy and trouble-free plant. It was rare in cultivation until recently (seed is now available). I grow it at the edge of a path with other pink-hued plants, including *Sedum pulchellum* and *Agastache* 'Cotton Candy', and with purple-flowered *Dalea purpurea*.

Phedimus selskianus

SYNONYM *Sedum selskianum*

A strikingly architectural plant with an open, arching, unbranched habit. Mature clumps may look like an octopus to some. Another unique feature is the overall hairiness of the plant parts (though, except for the silvered hair on the stems, a magnifying glass may be needed for confirmation). The stems are clothed in overlapping, spoon-shaped leaves like shingles on a roof; close inspection reveals that the leaves are toothed on their top halves. Normally matte-green, the foliage often turns bronze in the heat of summer. Late summer brings substantial flower clusters. These are composed of tiny, five-petaled blooms that sport bracts beneath and mimic the radiating habit of the stems. Deciduous. Propagate in spring by division, cuttings, or seed.

ZONES 4–9
PLANT SIZE 8 inches (20 cm) tall, 15+ inches (38+ cm) across.
SOIL dry to well-drained, average
LIGHT full sun
SIMILAR SPECIES AND CULTIVARS *Phedimus selskianus* 'Goldilocks' is a smaller form that can be grown from seed; *P. kamtschaticus*, and *P. sichotensis*.
ORIGIN from the border between China and the former Soviet Union.
LANDSCAPE AND DESIGN USES An extremely drought-tolerant plant, suitable for containers, rock gardens, and gravel gardens. It combines well with other strongly architectural plants such as *Euphorbia myrsinites*.

Phedimus sichotensis

SYNONYM *Sedum sichotense*

This plant looks just like a miniature form of *Phedimus kamtschaticus*. The best way to distinguish it, besides its smallness, is its more upright, clumping habit. The relatively thin leaves have a groove down their middle, notched tips, and typically measure around ¼ inch (6 mm) wide. As summer progresses, leaves can take on a purple-red hue; fall color is bright red. Stems are always dark red. In late summer, terminal flowers appear and are golden yellow. Propagate by cuttings.

ZONES 4–9, deciduous in Zone 5
PLANT SIZE 4 inches (10 cm) tall, 8–10+ inches (20–25+ cm) across.
SOIL dry to well-drained, average
LIGHT full sun
SIMILAR SPECIES AND CULTIVARS *Phedimus hybridus* 'Czar's Gold', *P. middendorffianus*, and *P. kamtschaticus* var. *floriferus* 'Weihenstephaner Gold'
ORIGIN Eastern Ukraine.
LANDSCAPE AND DESIGN USES Because of its smaller size, this plant requires a well-drained to dry site. It is very drought tolerant and a good choice for Southern gardens, where it is most likely evergreen. In the north, it dies back to a crown of green rosettes. Rare a few decades ago, it is now relatively available and is popular for planting en masse on greenroofs. I grow it with silver-leaved Missouri primrose (*Oenothera macrocarpa*) in a gravel garden.

Phedimus spurius 'Dr. John Creech'

SYNONYM *Sedum spurium* 'Dr. John Creech'

Once you've seen this sedum, it is hard to forget. The dense spreading mats of entirely green foliage are distinctive and the plant's best attribute. Take a closer look to appreciate how the foliage hugs and spoons with itself, leaving little room for anything else, including incursions of weeds. Summer brings bright pink flowers on 3- to 4-inch (7.5- to 10-cm) tall stems. Evergreen. Easily propagated in spring or summer from cuttings.

ZONES 3–9
PLANT SIZE 3–4 inches (7.5–10 cm) tall, 24+ inches (60+ cm) across.
SOIL dry to well-drained, average
LIGHT full sun to part shade
SIMILAR SPECIES AND CULTIVARS Frequently sold as 'John Creech'. *Phedimus spurius* 'Royal Pink' is quite similar, though it may be slightly looser in foliage. *Phedimus spurius* 'Roseum' foliage is similar to the species, though its habit is looser. It also has pink flowers, and its green foliage can take on red highlights. *Phedimus spurius* 'Heronswood Pink Stars' has strongly pointed light pink petals and darker pink carpels. *Phedimus spurius* 'Summer Glory' is a seed strain with dark pink flowers and a vigorous, but less substantial-looking habit.
ORIGIN garden origin
LANDSCAPE AND DESIGN USES One of the best for use as an evergreen groundcover due to its extremely tight mat and fast growth. It's a standout on greenroofs. The tight habit also makes it perfect along the edge of a walkway or garden.

Phedimus spurius 'Fuldaglut'

Fireglow two-row stonecrop
SYNONYM *Sedum spurium* 'Fuldaglut'

This excellent plant is basically an improved *Phedimus spurius* 'Schorbusser Blut' (dragon's blood). Compared to related plants, its foliage is much more substantial. The flat, rounded leaves with slight rounded teeth alternate in pairs across from each other, hence the common name of two-row stonecrop. The leaves appear more bronze than red due to green centers and red edges; if the plant gets full sun in summer, the color deepens to more red and less green. Its ruby flowers are also substantial, blooming on 5-inch (12-cm) stems that support open panicles or cymes 3–4 inches (7.5–10 cm) wide. In fall and winter, older leaves begin to fall off, leaving gray-brown stems tipped with rosettes of foliage. Semievergreen. Easily propagated from cuttings.

ZONES 3–9
PLANT SIZE 4–5 inches (10–13 cm) tall, up to 24 inches (60 cm) across.
SOIL dry to well-drained, average
LIGHT full sun to part shade
SIMILAR SPECIES AND CULTIVARS This variety is often sold under the English translation of its name, 'Fireglow' (or something similar). Its precursor *Phedimus spurius* 'Schorbusser Blut' (dragon's blood) has ruby-colored flowers but smaller foliage. It starts the growing season more on the green side, deepening to bronze red in summer, and its stems are thinner, only ⅛ inch (3 mm) wide. It won an Award of Garden Merit in 1993. 'Red Rock' is an all-red sport found by the author and introduced in 2013. Another similar plant is *Phedimus spurius* 'Bronze Carpet', which differs slightly in that its foliage tends to darken to bronze early and hold that color for most of the season; also, its flower color is lighter, more carmine.
ORIGIN garden origin; species native from Eastern Europe to Armenia and northern Iran.
LANDSCAPE AND DESIGN USES This plant is very durable and long-lived, so site it where you can enjoy its beauty for years to come. It looks fantastic with silver-leaved plants, such as lamb's ears (*Stachys byzantina*) and snow-in-summer (*Cerastium tomentosum*).

Phedimus spurius 'Leningrad White'

SYNONYM *Sedum spurium* 'Leningrad White'

This white form is vigorous and ground-hugging. In contrast to some of the other white-flowered varieties, it seems to have a fuller habit. Its all-green foliage can become pale in lean soil. White flowers, which appear in summer, are not usually numerous. The anthers are orange before they open, which adds a small touch of color. Evergreen. Propagate by division.

ZONES 4–9

PLANT SIZE 4 inches (10 cm) tall, 18–24+ inches (45–60+ cm) across.

SOIL dry to well-drained, average

LIGHT full sun to light shade

SIMILAR SPECIES AND CULTIVARS *Phedimus stoloniferus*, and other white *P. spurius* varieties—'Album' has strictly green foliage and white flowers that can be shy; somewhat less vigorous is 'Green Mantle' (synonym 'Album Superbum'), with paler foliage and white flowers that can be sporadic.

ORIGIN garden origin

LANDSCAPE AND DESIGN USES A great choice for greenroof or any massed or groundcover planting scheme, to break up the many pink, reds, and yellows that tend to dominate *Phedimus spurius* offerings. You can interchange this one with any of the similar varieties and get nearly the same look.

Phedimus spurius 'Red Carpet'

SYNONYM *Sedum spurium* 'Red Carpet', 'Atropurpureum', 'Coccineum', 'Elizabeth'

Beet-red foliage in combination with midpink flowers in summer distinguish this beautiful cultivar from similar plants. I have seen parts of the plant turn more bronze (a sport? This does not tend to affect the overall look in the garden, although the bronze areas may be a bit more vigorous). Also, in the colder months, the stems will persist with only a few leaves remaining at the tips. Semievergreen. Propagate by division or cuttings.

ZONES 4–9
PLANT SIZE 4+ inches (10+ cm) tall, 18–24+ inches (45–60+ cm) across.
SOIL dry to well-drained, average
LIGHT full sun to part shade

SIMILAR SPECIES AND CULTIVARS The softer pink flowers of the otherwise similar *Phedimus spurius* 'Raspberry Red' are a surprise, since the foliage is dark green to bronze with red highlights—a nice effect. *Phedimus spurius* 'Ruby Mantle' has larger foliage and ruby-colored flowers. The seed strain *Phedimus spurius* 'Voodoo' has become popular on greenroofs, probably as much for its name as for its dark red foliage; I have not found it to be vigorous or uniform in its dark color.

ORIGIN garden origin

LANDSCAPE AND DESIGN USES Pairing its red foliage with a gold-leaved sedum such as *Sedum makinoi* 'Ogon' is hard to beat. (Note that all the red-leaved forms are more similar in foliage than in flower.)

Phedimus spurius 'Tricolor'

Tricolor stonecrop

SYNONYM *Sedum spurium* 'Tricolor', *Phedimus spurius* 'Variegatum'

Aptly named. Most of the year, this handsome plant is easily identified by its white-edged foliage, which is tinged pink around green centers. At times, the pink is absent or not as pronounced. The white edge can vary as well, becoming quite thin, less than ⅛ inch (3 mm). Flowers are midpink. Semievergreen. Propagate by cuttings.

ZONES 5–9

PLANT SIZE 4 inches (10 cm) tall, 12–15+ inches (30–38+ cm) across.

SOIL dry to well-drained, average

LIGHT full sun

SIMILAR SPECIES AND CULTIVARS *Phedimus spurius* 'Fools Gold' looks so similar that it might be a synonym. *Phedimus kamtschaticus* 'Variegatus' has golden yellow flowers and its foliage has a creamy yellow variegated edge.

ORIGIN garden origin

LANDSCAPE AND DESIGN USES Be forewarned: this can plant revert and may need constant attention to keep the unwanted green shoots from outgrowing the variegated parts. It also seems to be more susceptible to late-spring frosts; when this happens, there is also a greater chance for reversion in the new growth. All that said, 'Tricolor' makes a great color contrast with any red-leaved sedum, such as *Phedimus spurius* 'Fuldaglut' (fireglow). It is justly popular in mixed containers.

Phedimus stoloniferus

Stolon stonecrop, narrow petal stonecrop

SYNONYM *Sedum stoloniferum*

It is a low, creeping plant whose stems root as they grow outward. The foliage is small, opposite, and remains green all year; the zigzagging stems are reddish. The light pink flowers are star-shaped. Evergreen. Previously this plant was rare in cultivation, but in recent years it has become available as seed. Propagated by cuttings, seed, or division.

ZONES 6–9

PLANT SIZE 2–3 inches (5–7.5 cm) tall, 12+ inches (30+ cm) across.

SOIL well-drained, average

LIGHT full to part shade

SIMILAR SPECIES AND CULTIVARS This plant is often confused with *Phedimus spurius*, but it is altogether smaller and not quite as hardy.

ORIGIN Iran to the Caucasus Mountains.

LANDSCAPE AND DESIGN USES Its tight, ground-hugging habit is appealing and would easily work where it could be walked on.

Phedimus takesimensis 'Golden Carpet'

SYNONYM *Sedum takesimensis* 'Golden Carpet'

Deep, glossy green color—at times, almost unreal in its vividness—and exceptional toughness distinguish this medium-size plant. The foliage is extra-thick, and both toothed and pointed. Thick, creeping stems are cinnamon brown and eventually turn upward. Substantial clusters of yellow flowers bloom in summer around the same time as *Phedimus kamtschaticus*. Evergreen. Plants may be self sterile; easy to propagate by cuttings.

ZONES 4–9

PLANT SIZE 6–8 inches (15–20 cm) tall, 8–10+ inches (20–25+ cm) across.

SOIL dry to well-drained, average

LIGHT full sun

SIMILAR SPECIES AND CULTIVARS *Phedimus takesimensis* is lighter green and has thinner, slightly longer foliage. Other similar plants include: *Phedimus aizoon, P. ellacombeanus,* and *P. hybridum*.

ORIGIN garden origin; species native to an island between Korea and Japan.

LANDSCAPE AND DESIGN USES This plant should be grown for its evergreen foliage. It makes a nice specimen in a container. It can also be used on a greenroof. Note that the foliage color can vary from deep green to pale, depending on soil fertility.

Prometheum sempervivoides

SYNONYM *Sedum sempervivoides, Rosularia sempervivoides*

An exotically appealing species that requires excellent drainage to prosper. Clusters of hairy, deep red flowers with yellow anthers (that, frankly, are too big for the base of the plant) cannot be missed when they bloom in early and midsummer. Typically these become so heavy that they start to fall over at some point. They also produce copious amounts of seed. Congested rosettes, meanwhile, feature pointy foliage; leaf color varies from green to almost all red. Out of bloom, the plants resemble slightly hairy hens and chicks, *Sempervivum*. Evergreen. Propagate by seed.

ZONES 3–9
PLANT SIZE 2–3+ inches (5–7.5+ cm) tall and wide, flower stems 6+ inches (15+ cm) tall.
SOIL rocky, dry to well-drained
LIGHT full sun

SIMILAR SPECIES AND CULTIVARS *Prometheum pilosum* (synonym *Sedum pilosum*) is much smaller, with thin, also-hairy foliage and pink flowers (it looks more like a *Sinocrassula* to me).
ORIGIN primarily from the Caucasus Mountains in Armenia.
LANDSCAPE AND DESIGN USES Because it dies after flowering, this small alpine plant should be planted in an area of superior drainage. (Wet winter conditions, in particular, are likely to be fatal.) An ideal site would be in fine gravel mulch, where the plants can reseed and, over time, form a colony. Your best bet would be to sow some of the seed right away and save some for the following year. In this way, there will be blooming plants every year instead of every other year. Natural companions are other rosette-formers, such as jovibarbas, rosularias, and sempervivums.

Rhodiola integrifolia

Entire leaved roseroot, king's crown, ledge stonecrop
SYNONYM *Sedum integrifolium*

An alpine beauty best enjoyed in areas similar to its native Rocky Mountain habitat. It has toothed foliage, whorled around the stems all the way to the tips where it holds the purple to red dioecious flowers (either male or female, on separate plants). The four stamens on the male flowers are longer than the thin petals; female flowers are also four-petaled. Deciduous. Propagate by division.

ZONES 3–9
PLANT SIZE 4–6 inches (10–15 cm) tall, 8–10 inches (20–25 cm) across.
SOIL rocky, well-drained, but moist and cool
LIGHT full sun
SIMILAR SPECIES AND CULTIVARS Its lookalike cousin *Rhodiola rosea* has yellow or red flowers, depending on where it is found. *Rhodiola integrifolia* var. *atropurpurea* has dark purple-red flowers on 16-inch (41-cm) tall plants; *R. integrifolia* subsp. *procera* has thinner, more glaucous foliage that is spear-shaped, ending in a point; *R. rhodantha* has pink flowers.
ORIGIN United States (Rocky Mountains and Alaska); British Columbia, Canada; Siberia.
LANDSCAPE AND DESIGN USES If you live in the Rockies or other cool, mountain climate, this is one to try. It makes a statement early in the season, fitting in nicely with many early-blooming alpine or rock garden plants. Because of its strict growing conditions, it is primarily sold in those areas where it native.

Rhodiola pachyclados

Afghani stonecrop, silver gem stonecrop
SYNONYM *Sedum pachyclados*

The tight mat of blue-green rosettes is a unique look. Thin rhizomes emerge near the mother plant, allowing the plants to creep. Foliage tips usually have three teeth and are tightly whorled around the end of the stems. Five-petaled flowers are white with slight yellow bases, accompanied by white ovaries and filaments. While the plant doesn't bloom heavily in the spring, it sometimes repeats in the fall. Although evergreen, it can look brown from the lower foliage that envelops the growth tip in winter. Propagate by cuttings, division, or seed.

ZONES 5–9
PLANT SIZE 4 inches (10 cm) tall, up to 18 inches (45 cm) across,
SOIL average to moist
LIGHT full sun
SIMILAR SPECIES AND CULTIVARS Although not related, there are other plants with blue-gray rosettes like *Sedum glaucophyllum*. Two varieties are in the trade—'White Diamond' and 'Nessy', which has more substantial, pointy foliage.
ORIGIN Afghanistan and Pakistan.
LANDSCAPE AND DESIGN USES The creeping, mat-forming nature of this plant makes it an exceptional groundcover. Growing conditions must be similar to its native mountain habitat—that is, places where temperatures are bound to stay cool most of the year. If your summers are hot and dry, provide irrigation or light shade or it will struggle.

Rhodiola rosea f. arctica

Roseroot, Arctic stonecrop
SYNONYM *Sedum roseum* f. *arcticum*

This attractive plant resembles euphorbia. It blooms early in spring when the blue-green foliage is still emerging. Individual leaves are wavy with a slightly toothed margin. Four-part chartreuse yellow buds are tightly packed at the tips and surrounded by leaves. Long yellow stamens stick out past the yellow petals, which have a strongly folded shape (male flowers have blunt, rounded tips). Below the tips, congested and whorled foliage extends all the way to the base. Deciduous. Propagate by division.

ZONES 1–6
PLANT SIZE 4–5 inches (10–13 cm) tall, 5–6 inches (13–15 cm) across.
SOIL well-drained, average to moist and rich
LIGHT full sun to part shade
SIMILAR SPECIES AND CULTIVARS This species varies widely in its native range and thus has many forms and subspecies. Mostly it is found in cultivation in two forms: the small form described above and a northern European form, *Rhodiola rosea*. That one reaches 18 inches (45 cm) or more tall and wide. It is usually dioecious, having male and female flowers on separate plants, but occasionally perfect flowers with both male and female parts on one plant are found. The North American forms are usually red-flowered, while the European/Asian forms are yellow.
ORIGIN widely distributed from Central Europe to Central Asia, Korea, Japan; also found in the eastern United States, Labrador, and Greenland.
LANDSCAPE AND DESIGN USES It looks terrific planted in a gravel garden next to other plants that appreciate good drainage, such as pinks (*Dianthus*) and Missouri primrose (*Oenothera macrocarpa*). It appreciates a cool location.

Sedum acre
Gold moss, wall pepper

The evergreen, ground-hugging nature of this plant, combined with the fact that it's super-easy to grow, make it is one of the most common of all sedums. Bright green spreading plants have congested triangular foliage whorled along thin stems. In summer yellow flowers ⅜ inch (9 mm) wide can be heavy enough to cover the plant. As the seedheads dry they become silver and woody, resembling driftwood in color. Evergreen. Easy to grow from seed or cuttings in spring or fall.

ZONES (2)3–8
PLANT SIZE 2–4 inches (5–10 cm) tall, 12 inches (30 cm) across.
SOIL rocky, dry to well-drained, average
LIGHT full sun to part shade
SIMILAR SPECIES AND CULTIVARS *Sedum acre* 'Oktoberfest' has white flowers. *Sedum acre* 'Minus' is one of the smaller forms, reaching only 1–2 inches. The large forms are related to the *S. acre* subsp. *majus*. *Sedum urvillei* has a similar look but tends to have some gray to its foliage; it also has quite a few variable forms and light yellow flowers. *Sedum nanifolium* is not cold-hardy but has similar-size leaves that hug the stem (amplexicaule), and the added interest of orange fall and winter color.
ORIGIN most of Europe; North Africa to Libya.
LANDSCAPE AND DESIGN USES Very adaptable and easy to grow. It can grow on a flat rock with an eastern exposure or on the side of the road. I have it planted on my mailbox in 2 inches (5 cm) of soil in part shade and it looks great. I've seen a simple planting of *Sedum acre* under white birch trees, which really complement each other. It is also a staple sedum on many greenroofs. Plants are readily available. Try growing the yellow-flowered species adjacent to its white-flowered cultivar 'Oktoberfest' as in the photo.

Sedum acre 'Aureum'
Gold moss stonecrop, gold wall pepper

A real winner—gold-tipped foliage on a vigorous evergreen plant. Tips measure less than ¼ inch (6 mm) wide with congested blunt foliage. This coloration is most prominent in spring, leading up to the early summer blooms. Yellow buds open to plentiful yellow flowers, giving the whole plant a golden appearance for up to two months. Evergreen. New plants can be started easily from cuttings in spring or fall.

ZONES 4–8
PLANT SIZE 2–4 inches (5–10 cm) tall, 12+ inches (30+ cm) across.
SOIL rocky, dry to well-drained, average
LIGHT full sun to part shade
SIMILAR SPECIES AND CULTIVARS *Sedum acre* 'Elegans' is a similar form with more silver or white new foliage in spring, leading into summer. *Sedum sexangulare* 'Golddigger' has chartreuse to gold whorled foliage.
ORIGIN garden origin
LANDSCAPE AND DESIGN USES Looks terrific in combination with similar-size green-leaved sedums. While this plant can suffer if summers are too hot, too humid, or too wet, some shade can help. Excessively wet soil in the heat of the summer or in winter also leads to decline.

Sedum adolphii

SYNONYM *Sedum nussbaumerianum*

A unique, eye-catching, tender plant. Thick succulent light green to orange foliage makes it stand out; in summer, the leaves may develop deeper colored margins. These leaves are ridged, alternate, and widest in the middle. In early spring, white starry flowers with pink anthers appear above green bracts, in full, rounded clusters. Individual flowers are ¾ inch (2 cm) across. The stems become woody in maturity. Winter color is a fleshy green. Not typically frost-hardy. Cuttings are easy.

ZONES 10–11
PLANT SIZE 6 inches (15 cm) tall, 18 inches (45 cm) across, hanging stems to 14 inches (36 cm) long.
SOIL dry to well-drained

LIGHT full sun
SIMILAR SPECIES AND CULTIVARS *Sedum adolphii* 'Coppertone' also has orange foliage in summer; it may be the same plant. Sometimes ×*Graptosedum* 'Golden Glow' (synonym *Sedum* 'Golden Glow') is sold as this plant, though its leaves are pale fleshy green. *Sedum adolphii* f. *variegata* is a rare variegated form.
ORIGIN Mexico
LANDSCAPE AND DESIGN USES A wonderful combiner, nice with so many other succulents, especially the blue and gray-green tropical rosettes of echeverias or the blue-fingered foliage of blue chalk sticks (*Senecio mandraliscae*). It can also be grown as a solo potted specimen.

Sedum album
Chubby fingers, white stonecrop

This is an attractive, if variable species. Foliage is finger-shaped and red to green. The amount of moisture appears to affect the coloration—in dry spells, the plants tend to darken to red or brown, while with ample moisture the plant will green up. White flowers appear in summer in relatively flat clusters; individual flowers are around ¼ inch (6 mm) wide, with five petals. Evergreen. Easy from seed, but some of the varieties will not come true; it's better to take cuttings or divisions in spring or fall.

ZONES 3–8, possibly colder
PLANT SIZE 4 inches (10 cm) tall, 16 inches (41 cm) across.
SOIL rocky, dry to well-drained
LIGHT full sun (tolerates part shade)

SIMILAR SPECIES AND CULTIVARS *Sedum album* 'Athoum' and 'France' are larger forms with more rounded foliage. *Sedum serpentini* is a relative with pink flowers that loses most of its leaves over the winter; it is sometimes listed as and may in fact be a subspecies of *S. album*. *Sedum stefco* has four-petaled flowers and foliage that turns bright red in winter and spring.

ORIGIN all European countries (except Ireland and Iceland), extending into North Africa.

LANDSCAPE AND DESIGN USES The ebb and flow of foliage colors from green to red is a good reason to use this in a spot where it will be viewed throughout the season. It roots from dropped leaves, making it a rapid spreader. In time, it forms mats well over 1 foot (30 cm) wide. Great in containers, fills in and hugs the upright plants next to it. A popular choice for greenroofs.

Sedum album 'Coral Carpet'▾

Coral Carpet stonecrop

This is a fast grower and has the additional virtue of uniform dark green to red-purple coloration. Cool weather renders it entirely red, a hue it holds for many months. Its growth habit tends to be more prostrate than the species. When it blooms, the plant becomes covered with clouds of white to pale pink flowers. Evergreen. Propagate by cuttings, in spring or fall.

ZONES 4–9
PLANT SIZE 4–6 inches (10–15 cm) tall, 27 inches (68 cm) across.
SOIL rocky, dry to well-drained
LIGHT full sun (tolerates light shade)
SIMILAR SPECIES AND CULTIVARS *Sedum album* f. *murale* has darker purple-red foliage and pink flowers (in the United States horticultural trade, it does not seem to be pink-flowered, so the true form may be lost).
ORIGIN garden origin
LANDSCAPE AND DESIGN USES A good groundcover and an excellent choice on greenroofs, thanks to its fast growth and handsome appearance. Best grown in lean or rocky soil to keep it from flowering itself to death. Removal of the seedheads after flowering helps ensure healthy, long-lived plants. (Decline can be attributed to excessive humidity or excessive moisture around the flower stems.)

Sedum album 'Fårö'▴

Small and adorable. It is the smallest form of *S. album*; like the species, it has contrasting green and red leaves throughout much of the year. Spring and summer color is primarily green; red highlights appear in cooler weather, then the foliage turns mostly red for the winter. The little leaves are not quite twice as long as wide at ⅛ inch (3 mm) wide and ¼ inch (6 mm) long). Evergreen. Propagate by division in spring or fall.

ZONES 4–9
PLANT SIZE 1–2 inches (2.5–5 cm) tall, 8 inches (20 cm) across.
SOIL dry to well-drained, average
LIGHT full sun
SIMILAR SPECIES AND CULTIVARS *Sedum serpentinii*, although its flowers are light pink, not white, and it loses most of its foliage in winter. Sometimes offered as *S.* ×*rubrotinctum* 'Mini Me', which is a misnomer.
ORIGIN Faro Island in Japan.
LANDSCAPE AND DESIGN USES Ideal for use with other extra-dwarf plants, alpines such as saxifrages, sempervivums, jovibarbas, phlox, silene, and thyme. Suitable for a rock garden, container, or terrarium. It can also be used in bonsai as an underplanting.

Sedum album var. micranthum 'Chloroticum'

SYNONYM *Sedum album*

A smaller plant notable for its entirely light green foliage (there are no red highlights). Intensity varies depending on soil fertility, however—in new, fertile ground, the leaves are darker green, while in lean soil the leaves become yellowish. Individual leaves can be nearly round. Evergreen. Increase by division.

ZONES 4–9
PLANT SIZE less than 1 inch (2.5 cm) tall, 10 inches (25 cm) across, flower stems 2–3 inches (5–7.5 cm) tall.
SOIL rocky, dry to well-drained
LIGHT full sun to light shade
SIMILAR SPECIES AND CULTIVARS *Sedum album* var. *micranthum* has similar small foliage but is not pale green. It has three distinct forms: 'Green Ice' stays olive green in summer and winter, 'Orange Ice' turns more orange over winter and in dry conditions, and 'Red Ice' is a reliable red form that greens up from spring into summer.
ORIGIN garden origin
LANDSCAPE AND DESIGN USES Due to its smaller size, it seems to be even more drought-tolerant than the species. Use it along hard surfaces where it can grow over and soften the edges. Also works as a groundcover in small spaces. Makes a pleasant contrast to other little sedums, including other forms of *Sedum album*, *S. acre*, and *S. dasyphyllum*. It would work well in topiaries as a base, setting off the large rosettes of succulents such as *Sempervivum*.

Sedum allantoides

A unique-looking plant. Light green to silver finger-shaped leaves arch toward the tip of the plant, getting smaller as they go up the stem. Flower stems are branched, with small white blooms in open groups. Evergreen. Take cuttings in spring or summer.

ZONES 10–11
PLANT SIZE 4–6 inches (10–15 cm) tall and wide.
SOIL well-drained
LIGHT full sun
SIMILAR SPECIES AND CULTIVARS *Sedum allantoides* 'Goldii' has wider foliage; the foliage of *S. pachyhyllum* is more blue-green.
ORIGIN Mexico
LANDSCAPE AND DESIGN USES Plants are vulnerable to cold weather and wet soil; definitely a plant for a hot, dry climate, or a container that can be brought inside for the winter. Try planting it with *Echeveria* and other tender broad-leaved succulents.

Sedum anglicum
English stonecrop

One of the tiniest sedums, its miniature leaves are barely 1/16 inch (1 mm) wide by 1/8 inch (3 mm) long. Mostly green, the foliage can have a reddish cast much of the year. The plant has a trailing habit and branched stems that eventually form a mat. In early summer, small white flowers appear on loose sprays; they may have a hint of pink. Evergreen. Raise from seed, cuttings, or division, best done in spring.

ZONES 5–9
PLANT SIZE 1–3 inches (2.5–7.5 cm) tall, 12–18 inches across (30–45 cm) across.
SOIL moist to average
LIGHT light shade (especially in the heat of summer)
SIMILAR SPECIES AND CULTIVARS *Sedum anglicum* 'Hartland' appears to be a more robust grower and bloomer; 'Minus' is a smaller form; 'Suzie Q' is a variegated sport of 'Hartland' with multicolored foliage of minty green, creamy white, and rosy pink that becomes more coral and blue-green in winter. Also similar are *S. acre* 'Minus' and *S. japonicum* var. *pumilum*.
ORIGIN Western Europe, primarily near the coast where there is high rainfall; Portugal to Great Britain, Ireland, and Scandinavia.
LANDSCAPE AND DESIGN USES Due to its small size, this is an ideal alpine or rock garden plant. It is also a great companion in a trough or other container with other small-statured plants. Plant it with other low growers like *Thymus praecox*. I recommend 'Hartland', if you can find it, for its size and vigor.

Sedum borschii

Cheerful yellow flowers against gray leaves make this species distinctive. The plants form congested mounds of smooth-textured, linear foliage (which is slightly ridged, or keeled, coming to a blunt tip). Relatively early in the season, the bright yellow flowers appear on red stems, which gives nice contrast. Individual flowers are five-petaled, 3/8 inch (9 mm) across. Evergreen. Can be increased by seed or division.

ZONES 4–9
PLANT SIZE 3–4 inches (7.5–10 cm) tall and wide.
SOIL rocky, dry to well-drained
LIGHT full sun
SIMILAR SPECIES AND CULTIVARS *Sedum borschii* 'Anna Schallach'. There is also plant in the trade sold as a sport of this species, but it has glossy foliage and white flowers, and is probably a crested form of *Sedum album*. *Sedum debile* is a rare, smaller relative with pale yellow flowers.
ORIGIN a small North American growing range—rocky sites in Idaho, Montana, Oregon, and Washington.
LANDSCAPE AND DESIGN USES It's best in a small container, crevice garden, gravel garden, or rock garden with other small cushion plants like *Dianthus* species. It will be more widely available in northern states and the Pacific Northwest.

Sedum brevifolium

Short-leaved stonecrop

A choice species treasured for its multicolored foliage. Thin stems are initially upright but in time sprawl into mounds, with the older foliage turning pink at the base to give the whole plant a three-colored effect of gray-green, chalky white, and pink. Leaves are rather stiff to the touch and are tiny, round, and carried in four (sometimes five) columns—another unique feature. They are covered with a whitish bloom (pruinose). The relatively few white flowers have five petals. Evergreen. Propagate by cuttings or division.

ZONES (5)6–9
PLANT SIZE 1–2 inches (2.5–5 cm) tall, 3–4 inches (7.5–10 cm) across.
SOIL rocky, dry to well-drained
LIGHT full sun
SIMILAR SPECIES AND CULTIVARS Can sometimes be confused with *Sedum dasyphyllum*, though the latter never has chalky foliage and its leaves are flat on top. Also, *S. brevifolium* is hard to the touch and its foliage does not drop off the stem with handling. There are unnamed forms with green foliage and red foliage. An attractive, more upright, slightly larger form is called *S. brevifolium* var. *quinquefarium*.
ORIGIN Corsica, France, Iberia, Sardinia, and North Africa.
LANDSCAPE AND DESIGN USES An ideal edging plant in containers. Typically slow-growing. Not overly common yet, but it deserves more use.

Sedum clavatum

A bit unusual for a sedum, somewhat reminiscent of an echeveria. It forms fat-leaved glaucous rosettes, more blue than green, with red-tipped leaves. Over time, plants grow more upright, exposing bare brown stems that eventually sprawl. Thick flower stems emerge from the tightly packed rosettes; when the stems reach 2–3 inches (5–7.5 cm) high, rounded flower clusters open. Sepals are a fleshy pink and may have red dots; a few leafy bracts are below. Individual flowers are white and five-petaled, with creamy white carpels and dark red stamens; overall, they fade to pink. The petals are wide at the base, ridged, and fold back at the base. Evergreen. Propagate by division or seed.

ZONES 10–11
PLANT SIZE 4 inches tall (10 cm), 8 inches (20 cm) across.
SOIL dry to well-drained, average
LIGHT full sun
SIMILAR SPECIES AND CULTIVARS *Sedum clavatum* 'Lime Drops' is a common lighter green form. *Sedum lucidum* has a similar size and habit but is green. *Sedum treleasei* has slightly thinner leaves and yellow flowers.
ORIGIN Mexico, specifically the valley along the Tiscalatengo River.
LANDSCAPE AND DESIGN USES In a mild climate (where freezing nights are rare), it can be used as a groundcover with other rosetted succulents like *Echeveria* or for defining a sharply drained edge of a wall. Elsewhere, it certainly makes a fine container plant (I have seen it successfully used in a hanging basket). It must be protected from freezing, especially in wet soil.

Sedum confusum

A longtime favorite. In the early 1900s, this plant was widely distributed. It's basically a branched subshrub, with most of the foliage on the top half of the plant. Leaves are shiny green and typically twice as long as wide, carried on short petioles. As they mature, their tips curl back. In late winter, loose panicles of yellow flowers appear. Individual flowers have six to seven petals, with uniform color in all parts, from petal to stamen to carpel. The stamens are extralong. Evergreen. Easy to propagate from cuttings.

ZONES (7)8–11
PLANT SIZE 5 inches (13 cm) tall, 12 inches (30 cm) across.
SOIL dry to well-drained
LIGHT full sun
SIMILAR SPECIES AND CULTIVARS
Sedum decumbens, also from Mexico, has smaller, thicker leaves that are wider at the base. *Sedum kimnachii* looks similar.
ORIGIN Mexico, specifically in the volcanic mountains of northern Puebla.
LANDSCAPE AND DESIGN USES
Although this species is from Mexico, it turns out to be semihardy. That is, it can actually freeze and recover. The main concerns are wet soil or repeated freeze and thaws. Its bright green foliage makes for a nice potted specimen. The plant easily makes the transition inside for the winter as a houseplant.

Sedum dasyphyllum

This unusual species produces a single, creamy white, wide-open flower that looks straight up from the top of the short plant. Tightly crowded evergreen foliage is typically pubescent (furry) and silver-gray, though in times of stress it takes on a pink or purple tinge. Leaves grow in spirals of four or five rows along the thin stems. When plants are disturbed or touched, the leaves fall off easily and can root to form new plants. Before the buds open, the back of the petals may be streaked pink, which gives the entire plant a soft pink cast. Small but numerous white flowers have wide, overlapping petals and creamy yellow ovaries. Evergreen. In spring, easily propagated by division or cuttings; may also be raised from seed.

ZONES 5–9
PLANT SIZE 1–2 inches (2.5–5 cm) tall, 6–8 inches (15–20 cm) across.
SOIL rocky, dry to well-drained

LIGHT full sun
SIMILAR SPECIES AND CULTIVARS *Sedum dasyphyllum* 'Atlas Mountain Form' has larger foliage compared to the species, and turns pink and silver with summer exposure; *S. dasyphyllum* 'Opaline' is larger and gray-green. *Sedum dasyphyllum* var. *macrophyllum* is more glaucous, with larger foliage (twice as big as the species); *S. dasyphyllum* subsp. *glanduliferum* is one of the extra-hairy (pubescent) forms.
ORIGIN Central Europe to the Mediterranean, south to North Africa from Spain and east to Turkey.
LANDSCAPE AND DESIGN USES In nature, the plant is found hanging on rocks and ledges, so try it in a comparable rock-garden setting or on a wall. Try it with other hanging plants such as purple rockcress. May also be used as a groundcover, but the site must have excellent drainage (excessive winter moisture causes losses). May also be used with bonsai.

Sedum dasyphyllum
var. *macrophyllum*

This botanical variety seems to be gaining in popularity for good reason, as it is very drought-tolerant. Fingered plants sprawl along the ground with white flowers held straight up. More silver than gray, the congested foliage is very attractive, resembling a smaller *Sedum morganianum* or *S. burrito*. Thick, substantial leaves, unlike the finer-textured, typically hairy species. Evergreen. Propagate by cuttings or division in spring or late summer.

ZONES 5–9
PLANT SIZE 3 inches (7.5 cm) tall, 10–12 inches (25–30 cm) across.

SOIL dry to average
LIGHT full sun to light shade
SIMILAR SPECIES AND CULTIVARS *Sedum dasyphyllum* 'Major' is most likely the same plant.
ORIGIN Central Europe to the Mediterranean, south into North Africa from Spain and east to Turkey.
LANDSCAPE AND DESIGN USES in the right climate (Zone 6 or warmer), it should make a good choice. I've seen it used on a living wall.

Sedum dendroideum
Tree sedum, tree stonecrop

This intriguing, tree-shaped, subshrubby sedum looks more like a jade plant (*Crassula*), to which it is distantly related. The "trunk" and stems are gray-brown and do look woodlike; the foliage is deep green and club-shaped. Look closely at the leaves—there is a line of windowlike glands along the leaf margin, which is unique on this species. Shy to flower, it blooms yellow in spring. Evergreen. Tender. Easy to propagate by cuttings.

ZONES 9–11(12)

PLANT SIZE 24 inches (60 cm) tall, 12 inches (30 cm) across.

SOIL dry to well-drained, average

LIGHT full sun

SIMILAR SPECIES AND CULTIVARS *Sedum confusum*, *S. decumbens*, and *S. praealtum* all have lighter green foliage.

ORIGIN Mexico south to Guatemala.

LANDSCAPE AND DESIGN USES Makes an excellent container plant, outdoors or inside. Due to its upright, open habit, it works well with other fillers and low-growing plants around it, such as *Sedum album* 'Coral Carpet'.

Sedum diffusum
Mouse-ear stonecrop

Given that it is one of the most common sedums in the wild in Mexico, it is no wonder that this species is being used in bordering states like Texas, New Mexico, Arizona, and California as well as other southern states. Plants are open and sprawling, with leaves ranging from pale green to gray-green. Pink highlights brighten up the older foliage, adding dimension. Leaves are small, thin, fingered, and hold themselves close to the stems. White flowers bloom at the tips in summer. Evergreen. Easy to propagate from cuttings.

ZONES 7–11
PLANT SIZE 2+ inches (5+ cm) tall, 10–12+ inches (25–30+ cm) across.
SOIL rocky, dry to well-drained
LIGHT full sun to light shade

SIMILAR SPECIES AND CULTIVARS *Sedum diffusum* 'Potosinum' is larger and more vigorous; its habit is more compact, however, and the foliage is more silver.
ORIGIN Mexico
LANDSCAPE AND DESIGN USES Excellent both in containers and in the garden. I saw one in a container in Texas hanging down 6–8 inches (15–20 cm), with air roots coming out of the stems, ready to fall to the ground and grow. I brought a cutting home in an envelope, but forgot about it; when I remembered, a full month later, I was amazed that it was still alive. This is a testament to how tough this plant can be and how long this sedums (and others) can go without water.

Sedum divergens

Cascade stonecrop, old man's bones
SYNONYM *Sedum globosum, Amerosedum divergens*

Especially glossy evergreen foliage and a nearly round leaf shape distinguish this handsome species. Close inspection reveals that the leaves grow in pairs, opposite to each other and at right angles to the following pair. Foliage is usually red and green at the same time, though more sun brings out the red. The plant can be shy to flower, but when it does, the blooms appear in late spring or early summer and are yellow; individual flowers are six-petaled and strongly ridged down the middle. Evergreen. Propagate from seed or stem cuttings.

ZONES 5–9

PLANT SIZE 6–8 inches (15–20 cm) tall, 8–12+ inches (20–30+ cm) across.

SOIL rocky, moist

LIGHT full sun to some shade

SIMILAR SPECIES AND CULTIVARS There are smaller and larger forms. The smaller tend to come from higher altitudes. One of these, *Sedum divergens* var. *minus*, is dark green and has a fuller habit. *Sedum oreganum* is smaller. *Sedum debile* is a petite alpine species from the Rocky Mountains; it is pale green and somewhat rare in cultivation.

ORIGIN Pacific Coast of North America.

LANDSCAPE AND DESIGN USES This species prefers cool summers and is known to tolerate temporarily dry conditions. Due to its evergreen nature and the fact that it is native to the area, it is a fine choice for Pacific Northwest gardens and greenroofs. Try it in combination with *Sedum oreganum* or *S. spathulifolium*.

Sedum emarginatum 'Eco-Mt. Emei'

Mt. Emei stonecrop

SYNONYM *Sedum makinoi* var. *emarginatum* 'Eco-Mt. Emei'

The best way to identify this perky little Chinese groundcover is by the telltale notch at the top of each nearly round, dark green leaf. The leaves are carried on petioles and are opposite. In summer, gold flowers are spread out on the plant, rarely touching or overlapping. Winter foliage is reddish. Evergreen. Easy to propagate by cuttings.

ZONES (5)6–10
PLANT SIZE 4–6 inches (10–15 cm) tall, 24 inches (60 cm) across.
SOIL dry to well-drained, average
LIGHT full sun to part shade
SIMILAR SPECIES AND CULTIVARS *Sedum emarginatum* is lighter green and less vigorous. Strongly resembles the *Sedum makinoi* types; the main difference is the notched leaves. *Sedum makinoi* 'Salsa Verde' is smaller, dark green, with notched leaves.
ORIGIN China; collected and introduced by Don Jacobs of the now-defunct Eco-Gardens nursery in Georgia.
LANDSCAPE AND DESIGN USES Ideal as a quick groundcover for partly shaded, moist sites. In hot, humid climates, this plant can be used on a greenroof if given some shade. It combines well with other sedums of similar growth habit and appearance, such as *Sedum makinoi* and its cultivars, and *S. tetractinum*.

Sedum furfuraceum

Intriguing foliage! The color is dark green but with purple hues on a cracked-shell-like surface; leaves are egg-shaped. Stems don't grow straight, which gives the whole plant a bushy appearance. The flowers are whitish, but can gain hints of pink depending on the exposure. Relatively slow-growing. Evergreen. Propagate by cuttings.

ZONES 10–11
PLANT SIZE 10–12 inches (25–30 cm) tall, 12 inches (30 cm) across.
SOIL dry, well-drained
LIGHT full sun
SIMILAR SPECIES AND CULTIVARS *Sedum hernandezii* is brighter green. *Sedum* 'Crocodile' has dark green, glossy, long, jellybean-like foliage.
ORIGIN Mexico.
LANDSCAPE AND DESIGN USES This is definitely best as a container plant, due to its slow growth rate and its compact, bushy habit. For some contrast, try it with *Sedum hernandezii* or *S. pachyphyllum*.

Sedum glaucophyllum

Glaucous stonecrop, cliff stonecrop

Easily recognized by its tight, spreading rosettes with smooth (entire) foliage of glaucous silver. The flowers start out white and fade to pink; they have four petals and flat filaments (the part that holds the reddish brown anthers) fixed to the bases of the petals. Evergreen. Easy to propagate by division, but it comes true from seed.

ZONES (5)6–9
PLANT SIZE 4 inches (10 cm) tall, 6–8 inches (15–20 cm) across.
SOIL rocky, average to moist
LIGHT part shade
SIMILAR SPECIES AND CULTIVARS *Sedum glaucophyllum* 'Silver Frost', which is sometimes sold as *S. nevii* (a misnomer), has lighter green rosettes, with longer leaves that have toothed tips. I propose calling the darker, more green form 'Red Frost' to set the two distinct forms apart and account for the red hues the plant takes on in cool weather.
ORIGIN Southeastern United States, in the Appalachian Mountains.
LANDSCAPE AND DESIGN USES Plant this one in some shade. It does not tolerate excessive dryness. A great plant for hot, humid, coastal regions, along with its relatives *Sedum ternatum* and *S. pulchellum*.

Sedum gracile ▲

This one looks a like a petite version of *Sedum sexangulare*. It, too, has a branching and sprawling habit. Foliage is green and linear; congested tips swirl above the lower stems, which typically have some dried foliage hanging on. In summer, stemless white flowers bloom at the tips in small clusters. Evergreen. Propagate by division.

ZONES 5–8
PLANT SIZE 2 inches (5 cm) tall, 6 inches (15 cm) across.
SOIL rocky, dry to well-drained
LIGHT full sun
SIMILAR SPECIES AND CULTIVARS *Sedum grisebachii* and *S. sexangulare*, although both of these have yellow flowers.
ORIGIN Caucasus Mountains, in alpine and subalpine areas.
LANDSCAPE AND DESIGN USES This mountainous plant does not like hot conditions. That said, it is a good choice in a small area. Combine it with other little alpines like drabas and jovibarbas. Where summers are not too hot, it may be used on a greenroof.

Sedum grisebachii

SYNONYM *Sedum kostovii*

Get out your magnifying glass. This form has tiny little nipples at the ends of its already small foliage. Crowded leaves on tightly mounding plants are another attribute that set it apart from similar spreading species. In summer, the foliage gains a rusty red cast, then quickly gets covered by greenish yellow flowers only around ¼ inch (6 mm) wide. Evergreen. Propagate in spring or fall by cuttings or division.

ZONES 5–9
PLANT SIZE 3–4 inches (7.5–10 cm) tall, 4–5 inches (10–13 cm) across.
SOIL dry to well-drained, average
LIGHT full sun
SIMILAR SPECIES AND CULTIVARS *Sedum sexangulare* is spreading, not clumping; *S. laconicum* is similar but a bit larger, with pale yellow flowers.
ORIGIN the perennial form hails from high altitudes in Greece and Bulgaria.
LANDSCAPE AND DESIGN USES The rounded habit and small stature of this plant make it perfect for troughs and other small containers. It combines well with small alpine species of *Campanula* and *Dianthus*.

Sedum gypsicola

SYNONYM *Oreosedum gypsicola*

Closely related to *Sedum album*, but its leaves are somewhat flat on top, with a closely overlapping habit at the tips. Also, the leaves are often downy or have small bumps on their usually dark surface. The foliage spirals slightly along the stems. The plant blooms in open cymes in early summer on tall, upright stems. The pretty flowers are white with maroon anthers. Evergreen. Easy to propagate by cuttings or seed.

ZONES 6–9
PLANT SIZE 2–3 inches (5–7.5 cm) tall, 6–8 inches (15–20 cm) across, flower stems 6–8 inches (15–20 cm) tall.
SOIL rocky, dry to well-drained
LIGHT full sun
SIMILAR SPECIES AND CULTIVARS *Sedum album*
ORIGIN Spain and the mountains of North Africa.
LANDSCAPE AND DESIGN USES This form is not overly common in the United States. Due to its southern European origins, it should make a good candidate for greenroofs in hot, humid regions. It makes a nice companion to *Saxifraga* and *Talinum calycinum*, which also carry their flowers on wiry stems.

Sedum hernandezii ◄

Jellybean plant

A deserved favorite. Thick, fingered foliage emerges bright green but becomes darker; it can have an unmistakable cracked texture (some might call it scaly), especially in the older leaves. Upright hairy stems hold the 1-inch (2.5-cm) foliage in four columns. In late winter to spring, yellow flowers bloom in small clusters at the top of the short, structured plants. As plants mature, they become bushy. Evergreen. Easy to propagate from leaf cuttings.

ZONES 10–11
PLANT SIZE 4 inches (10 cm) tall, 12 inches (30 cm) across.
SOIL well-drained
LIGHT full sun to part shade
SIMILAR SPECIES AND CULTIVARS *Sedum furfuraceum, S.* ×*rubrotinctum*
ORIGIN Mexico
LANDSCAPE AND DESIGN USES This makes a unique houseplant. Try it combined with other jellybean types like *Sedum stahlii* and *S.* ×*rubrotinctum*.

Sedum hispanicum var. *hispanicum*

Spanish stonecrop

A handsome little spreading cushion plant. Its foliage is more purple and its habit is tighter than the silvery but otherwise similar *Sedum hispanicum* var. *minus*. Also, unfortunately, this one is not quite as perennial or hardy. The flowers are white with pink stripes. Evergreen. Propagate by division.

ZONES (5)6–9

PLANT SIZE 4–5 inches (10–13 cm) tall, 12+ inches (30+ cm) across.

SOIL dry to well-drained, average

LIGHT full sun

SIMILAR SPECIES AND CULTIVARS More commonly sold as a cultivar named 'Purple Form' or 'Purpureum'. This plant can be confused with *Sedum pallidum*, which is rare in cultivation and acts more as an annual; it has extra-large white inflorescences on arching stems. *Sedum pallidum* var. *bithynicum* is short-lived but can be hardy.

ORIGIN Sicily (Italy) to Greece, Turkey, into the Middle East.

LANDSCAPE AND DESIGN USES I have always enjoyed growing this with the more silvery *Sedum hispanicum* var. *minus* for the contrast in foliage color.

Sedum hispanicum var. *minus*

Tiny buttons
SYNONYM *Sedum hispanicum* subsp. *glaucum*

A variable species, although fine-textured blue-green foliage on spreading to cushion-forming plants is common. The flowers look pink but are actually creamy white with a pink stripe down the middle; the backs also have the pink stripe, along with tiny hairs blending into the pink flowering stem (peduncle). Petal count varies from four to 10. Evergreen. Easy to propagate by division in spring or fall; at the nursery, we raise it over the winter in a cool but heated greenhouse.

ZONES 5–9
PLANT SIZE 4–5 inches (10–13 cm) tall, 20 inches (50 cm) across.
SOIL gravelly, dry to well-drained, average
LIGHT full sun
SIMILAR SPECIES AND CULTIVARS *Sedum hispanicum* var. *hispanicum* and *S. hispanicum* var. *minus* 'Aureum', although neither is as vigorous or hardy. There seems to be some confusion with *S. pallidum* var. *bithynicum*; however, that is a white-flowered form, both hardier and short-lived.
ORIGIN Sicily (Italy) to Greece, Turkey, into the Middle East.
LANDSCAPE AND DESIGN USES In areas with wet winters or in wet soil, this plant is not reliable over the winter, but in the right spot and with good drainage, it is perennial in Zone 5. The fine texture blends nicely especially with other fine-textured choices, including *Sedum acre*, *S. album*, *S. dasyphyllum*, and *S. sexangulare*. All are great container plants, supplying contrasting texture to the broad foliage of other plants.

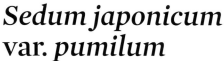

Sedum japonicum var. *pumilum*

SYNONYM *Sedum oryzifolium* 'Tiny Form'

This is one of the very smallest sedums in cultivation. The foliage color is unique; in winter it is orange to bronze, and in summer it greens up while still retaining some orange or bronze highlights or speckling. The creeping stems hug the ground. The fact that the plant is hardy is surprising. Evergreen. Propagate by division.

ZONES 5–9
PLANT SIZE 1 inch (2.5 cm) tall, 4 inches (10 cm) across.
SOIL well-drained
LIGHT full sun
SIMILAR SPECIES AND CULTIVARS *Sedum oryzifolium* is more upright in habit but is rare in the United States.
ORIGIN coastal islands of Japan and other nearby Pacific islands.
LANDSCAPE AND DESIGN USES Its creeping habit makes this one ideal for planting in small cracks between stones and in walkways between pavers. Although it is not common, it deserves more use. I mix it with *Sedum dasyphyllum*; it would also work with *S. gracile* and the small forms of *S. album*.

Sedum japonicum var. *senanese*

Japanese stonecrop

Quite attractive! A loose sprawling plant with deep green foliage enhanced by red highlights for most of the year. The leaves are dull-pointed and flat on both sides. The flowers, which appear in small numbers at the growing tips, are yellow. Evergreen. Propagate by cuttings.

ZONES 6–9
PLANT SIZE 2–3 inches (5–7.5 cm) tall, 4–5 inches (10–13 cm) across.
SOIL moist but well-drained
LIGHT part shade
SIMILAR SPECIES AND CULTIVARS *Sedum japonicum* has smaller, lighter green foliage and a smaller stature. *Rhodiola trollii* (synonym *Sedum trollii*) resembles this, and quite a few other sedums from the same region.
ORIGIN Honshu Island, Japan; Taiwan.
LANDSCAPE AND DESIGN USES The orange-red winter highlights will make this ground-hugger stand out. Would look good with *Sedum nokoense*.

Sedum japonicum 'Tokyo Sun'

Golden Japanese stonecrop

Fine-textured, loose mounds have some of the brightest foliage of all the gold sedums. It is quickly becoming more widely available despite not being fully cold-hardy. As cooler weather approaches, its coloration fades and moderates to green with orange-brown highlights, eventually turning bronze for the winter. The growth habit is prostrate, and the leaves are flat on both sides. This plant rarely blooms, but reportedly has yellow flowers. Evergreen. Propagate by cuttings or division.

ZONES 7–11

PLANT SIZE 2 inches (5 cm) tall, 4–5 inches (10–13 cm) across.

SOIL well-drained to average, moist

LIGHT full sun to part shade

SIMILAR SPECIES AND CULTIVARS I saw this plant labeled *Sedum japonicum* f. *leucanthemum* at ItSaul Plants in 2007. While I have not seen the name 'Tokyo Sun' in print a lot, I believe it makes sense to use it as a valid cultivar name instead of the descriptive "Fine Yellow Form"; it appears to be in use in Japan. *Sedum japonicum* var. *pumilum* is a related miniature that is more cold-hardy (to Zone 5).

ORIGIN garden origin

LANDSCAPE AND DESIGN USES Be vigilant—the gold can revert to green (however, the green form, the species, is not commonly sold). This is an adaptable plant, provided it gets some shade and does not freeze. It would make a great addition to mixed containers, playing off the other plants' colors and textures; try it with *Sedum lydium* or *S. dasyphyllum*.

Sedum kimnachii

This subshrub eventually develops a perennial but not woody stem. Shiny, bright green, spoon-shaped foliage becomes more congested at the top of the plant. Winter color tends to orange. Meanwhile, deep yellow to golden yellow flowers are borne on loose, branched heads in late winter to early spring. Before they open, you can observe large, irregular green sepals; these peek through the sides of the five large petals later when the flowers are fully open. Evergreen. Propagate by cuttings.

ZONES (7)8–11
PLANT SIZE 4 inches (10 cm) tall, 10 inches (25 cm) across.
SOIL dry to well-drained
LIGHT full sun
SIMILAR SPECIES AND CULTIVARS *Sedum confusum*, *S. decumbens*
ORIGIN Mexico
LANDSCAPE AND DESIGN USES Still not widely available, but hopefully that will change. While this plant might look like it belongs in the house, it does surprisingly well outside, even tolerating freezing temperatures provided the soil is not too wet. Use as an accent with hardy succulents like agaves, aeoniums, or aloes.

Sedum lanceolatum

Lance-leaf stonecrop
SYNONYM *Petrosedum lanceolatum*

According to sedum-expert Ray Stephenson, this plant is a biennial. I believe the plant being sold in the United States horticulture trade and by Jelitto Seeds these days, however, is a perennial. From first-year rosettes, creeping stems of light gray-green to plum-brown root easily. Needlelike gray-green foliage develops purple coloration in more sun and may turn red in winter. Leaves are more crowded at the tips, more sparse toward the base of the stems. If it blooms, flat clustered buds on 4-inch (10-cm) stems open to yellow flowers. Evergreen. Easy to propagate by cuttings

ZONES 4–9
PLANT SIZE 4 inches (10 cm) tall, 8 inches (20 cm) across, flower stems 6–8 inches (15–20 cm) tall.
SOIL rocky, well-drained to dry
LIGHT light shade (tolerates full sun)
SIMILAR SPECIES AND CULTIVARS There are many forms and subspecies in the wild; this plant has a wide range for a native stonecrop. Some of the forms are much tighter, almost clumping. *Sedum lanceolatum* subsp. *alpinum* is a smaller form from alpine sites. *Sedum lanceolatum* var. *nesioticum* is a larger form that grows from British Columbia into Washington close to sea level. *Sedum lanceolatum* var. *rupicola*, from the Wenatchee Mountains (a spur of the Cascades) in Washington State, has small green foliage and red stems. Sometimes nurseries confuse this plant with *S. stenopetalum*, another North American native that grows in a similar habitat; however, that one has green foliage with red highlights. To my eye, this plant most closely resembles a more compact form of *Petrosedum rupestre* 'Blue Spruce'.

ORIGIN British Columbia south into the Rockies and east into the Great Plains.

LANDSCAPE AND DESIGN USES This would make a nice companion to other native clumpers like buckwheat and rock jasmine.

Sedum laxum
Roseflower stonecrop

Recognizable by its loose rosettes of extra-thick leaves up to ¼ inch (6 mm) deep. These are arranged crosswise and opposite from each other (decussate). Some of the leaves can have a notch at the end instead of a tip. The pretty pink flowers have their petals fused halfway up, which is not typical of sedum; anthers are red. Semievergreen. Propagate by cuttings.

ZONES 6–9
PLANT SIZE 2–3 inches (5–7.5 cm) tall, 5–6+ inches (13–15+ cm) across.
SOIL rocky, sharp drainage
LIGHT full sun to light shade
SIMILAR SPECIES AND CULTIVARS There are many subspecies. *Sedum laxum* subsp. *laxum* has spoon-shaped foliage and thin leaves along the flower stem; flowers can be white to creamy yellow. *Sedum laxum* subsp. *eastwoodiae* has smaller glaucous foliage only ⅜ inch (9 mm) long, and leafy stems. *Sedum laxum* subsp. *heckneri* has reflexed buds above fingernail-shaped leaves that clasp the stem; its flowers have yellow anthers, compared to the typical red. *Sedum laxum* subsp. *latifolium* has the best vigor and also has larger, greener foliage both in the basal rosettes and on the flowering stem. *Sedum oregonense* and *S. obtusatum* have larger blue foliage and are choice forms.
ORIGIN Southern Oregon and Northern California.
LANDSCAPE AND DESIGN USES Planting it with other Pacific Northwest native plants is ideal. Try it with *Lewisia* and other sedums such as *Sedum oreganum* and *S. divergens*.

Sedum lineare 'Variegatum'
Needle stonecrop

This one's foliage really grabs attention. The leaves are light green and linear, with a thin white edge; they are flat and whorled in rows of three. The plant's habit is initially an upright tuft, then sprawling. Flowering is fairly rare, but when it occurs, the flowers are yellow. In fall, the stems tend to become brittle and break off easily, but there's usually a new rosette at soil level to resume growth the following year. Semievergreen, but not hardy. Easy from cuttings.

ZONES 9–11
PLANT SIZE 5–6 inches (13–15 cm) tall, 15+ inches (38+ cm) across.
SOIL rocky, dry to well-drained, average
LIGHT full sun to light shade
SIMILAR SPECIES AND CULTIVARS *Sedum mexicanum*, *S. sarmentosum*
ORIGIN Honshu Island, Japan; possibly China.
LANDSCAPE AND DESIGN USES Its sprawling habit makes this a perfect plant for a spot where it can overhang an edge—a hanging basket, or a wall. In cold climates, bring it in for the winter; alternatively, take a cutting with the intention of replanting the following spring after the last frost. This sedum also can be grown as a houseplant.

Sedum lucidum

Thick glossy foliage varies in color from an almost olive green to pale green-yellow at the base but almost always having some red to orange-coral tips. Somewhat unusual is the fact that as the stems elongate, the foliage holds onto the stems. White flowers emerge close to the tips in early spring on upright curved pedicels that thicken as they get closer to the bud. Round clusters of white flowers have greenish to creamy yellow carpels that change color in time, becoming coral-orange like the leaf tips. Evergreen. Propagate by division or cuttings.

ZONES 10–11
PLANT SIZE 9 inches (23 cm) tall, 5–6 inches (13–15 cm) across.
SOIL rocky, dry to well-drained
LIGHT full sun
SIMILAR SPECIES AND CULTIVARS
There is a crested form, but it is rare. *Sedum lucidum* 'Obesum' has glossier foliage. ×*Graptosedum* (synonym *S.* 'Golden Glow') and *S. clavatum* are also similar.
ORIGIN Mexico
LANDSCAPE AND DESIGN USES
This is an easy container plant, providing lots of contrast all by itself, but I recommend planting it with other thick-rosette plants such as *Sedum adolphii*, *S. clavatum*, *S. treleasei*, and echeverias.

Sedum ×luteoviride

The most notable feature of this subshrub is its thick short leaves at right angles to the stem. Both the stem and the leaf tips typically have fleshy red coloration bleeding into the green bases, which can deepen to orange-red in full sun. Winter color tends to cinnamon with green. Cheery yellow flowers appear in profusion early in the year, in both terminal and lateral clusters, with smaller, bractlike leaves tucked in close to them. Over time, the plant becomes dense, with the lower foliage falling away from the stem. Evergreen. Easy from cuttings.

ZONES (8)9–11

PLANT SIZE 6 inches (15 cm) tall, 5–6+ inches (13–15+ cm) across.

SOIL dry to well-drained, average

LIGHT full sun

SIMILAR SPECIES AND CULTIVARS Both parents are very different from this hybrid: *Sedum praealtum* has long, entire, flat leaves, while *S. greggii* has small, cone-shaped rosettes of foliage. Other comparable plants are *S. decumbens*, *S.* 'Little Gem', *S. reptans*, and *S.* 'Rockery Challenger'.

ORIGIN a natural hybrid between *Sedum praealtum* subsp. *parvifolium* and *S. greggii*, from Mexico.

LANDSCAPE AND DESIGN USES This is a terrific plant for a small container. It needs bright light over winter. In a warmer climate, it could be tried outside, but I would keep an extra plant inside for winter, just in case.

Sedum lydium

Mossy stonecrop

The unusual color always stands out; ball-shaped, bright green leaves tipped in red are dramatic against red stems. The entire plant can turn entirely red in drought and in colder times of the year (spring and fall). The stems of this spreading, creeping, tufted form are mostly bare below the tips. Very shy to bloom, but if they appear, the flowers are bicolor pink-and-white. Evergreen. Propagate from cuttings.

ZONES (5)6–8
PLANT SIZE 1–2 inches (2.5–5 cm) tall, 6–8 inches (15–20 cm) across.
SOIL moist, but with sharp drainage
LIGHT light to part shade
SIMILAR SPECIES AND CULTIVARS This one doesn't have any very close relatives, although the buttonlike foliage and clumping habit is reminiscent of *Sedum grisebachii*.
ORIGIN Turkey to Armenia.
LANDSCAPE AND DESIGN USES Pair it with other similarly fine-textured species like *Sedum hispanicum*, the gold-leaved *S. japonicum* 'Tokyo Sun', or *S. sexangulare* 'Golddigger'. It can be grown in areas with some foot traffic, like between stepping stones and other walking-path materials.

Sedum makinoi
Makino stonecrop

A creeping species whose growth habit changes with the seasons. In winter, the tips become dense with rounded leaves; in summer, the stems elongate and the spoon-shaped leaves are more widely spaced. Blooming occurs in early summer and can be sparse. Carried on a zig-zagging stem, the single green-yellow flower is backed with leafy bracts and set apart from any adjacent flowers. Evergreen but not fully hardy. Propagate by division in fall or winter.

ZONES 7–9
PLANT SIZE 2 inches (5 cm) tall, 8–12 inches (20–30 cm) across.
SOIL rocky, wet to moist
LIGHT part sun to shade
SIMILAR SPECIES AND CULTIVARS *Sedum makinoi* 'Lime-light' is a lighter green compared to the species, and at times chartreuse; 'Ogon' is the gold-leaved form; 'Salsa Verde' has smaller, notched foliage that turns orange-red in cool weather; 'Variegatum' has a thin white edge. *Sedum tetractinum* is also similar.
ORIGIN Japanese islands
LANDSCAPE AND DESIGN USES This species has become more widely grown, with both the variegated form and an-all gold form rising in popularity at the same time. It makes a fine groundcover in hot, humid climates.

Sedum makinoi 'Ogon'
Golden Makino stonecrop

A superb plant. Foliage varies from gold to yellow, depending on the time of year. The plant is shy to bloom, but if it does, the flowers appear in early summer and are yellow with leafy yellow bracts just below. The habit changes over the course of the growing year, too; in winter, the mostly round, butter-colored leaves are crowded, while in summer, the plant spreads out and the leaves are more spoon-shaped. Evergreen. Propagate by cuttings.

ZONES 7–9
PLANT SIZE 2 inches (5 cm) tall, 8–12 inches (20–30 cm) across.
SOIL rocky, wet to moist
LIGHT part sun to shade
SIMILAR SPECIES AND CULTIVARS *Sedum makinoi* 'Lime-light' and 'Variegatum'. 'Ogon' and 'Variegatum' were brought to the United States by plantsman Dan Heims of Terra Nova Nurseries around 1989; he traded a fellow grower in Japan.
ORIGIN garden origin, the species is from Japan.
LANDSCAPE AND DESIGN USES This plant is excellent in combinations. It is especially stunning with dark-leaved plants, for instance, black mondo grass (in Zone 6 or warmer). It also pairs easily with the blue foliage of *Hylotelephium sieboldii*, *H. cauticola*, and *H. ewersii*. I prefer this plant's winter aspect, when its growth habit becomes more crowded and its foliage color mellows. Fall is a good time to start propagating more plants for spring, as the plant becomes more brittle then and stems break off easily.

Sedum mexicanum

Golden ball stonecrop,
Mexican stonecrop
SYNONYM *Sedum mexicanum* 'Golden Ball'

A bright, cheerful-looking plant. Fresh green linear leaves whorl, mostly in sets of four, along upright then sprawling stems. Leaves have spurs at the bases. Stem color is light green to palest pink (a somewhat unusual characteristic only seen in a few of the related species). Flowers, if they appear, occur on thick, branched stems, are five-petaled and yellow with orange-red anthers. As the plant goes from fall to winter, old stems tend to become brittle and break off; usually there are some new rosettes at ground level to resume growth in spring. Evergreen in frost-free winters. Easy from cuttings in spring or fall.

ZONES (7)8–11
PLANT SIZE 6–8 inches (15–20 cm) tall, 15+ inches (38+ cm) across.
SOIL well-drained to moist
LIGHT part shade to nearly full shade; with moisture or irrigation, tolerates full sun.
SIMILAR SPECIES AND CULTIVARS This species looks similar to *Petrosedum rupestre* and its variations, and is often mistaken for a hardy *P. rupestre*. *Sedum lineare* and *S. sarmentosum* are similar. *Sedum* 'Rockery Challenger' is thought to be a hybrid of this species and *S. sarmentosum*.
ORIGIN not clear; according to American nurseryman Tony Avent, it may have been imported from Southeast Asia.
LANDSCAPE AND DESIGN USES Its upright yet rapidly spreading nature make it a good choice where it can hang over an edge of a container or wall. It is also ideal when a seasonal display is needed quickly.

Sedum mexicanum 'Lemon Ball'

SYNONYM *Sedum* 'Lemon Coral'

You can't miss this one—its whorls of bright gold linear leaves and matching stems form a bushy, ball-shaped plant that is very distinctive. The stems are upright but eventually sprawl. The plant is a fast grower. It is not a reliable bloomer, though if it does, they will be yellow flowers on 8-inch (20-cm) tall stems. Evergreen if protected from freezing. Easy from cuttings.

ZONES 8–11
PLANT SIZE 6–8 inches (15–20 cm) tall, 15+ inches (38+ cm) across.
SOIL well-drained to moist
LIGHT part to nearly full shade; with moisture or irrigation, tolerates full sun
SIMILAR SPECIES AND CULTIVARS The popular *Petrosedum rupestre* 'Angelina' also has golden foliage, but develops orange and copper-red winter highlights. Two other similar choices are *Sedum lineare* and *S. sarmentosum*.
ORIGIN garden origin
LANDSCAPE AND DESIGN USES Makes a unique hanging plant, and does well in the garden in frost-free areas. Tends to remain the same color year-round. Recently made more popular by Proven Winners, but they originally listed it as the hardy species *Petrosedum rupestre*.

Sedum mocinianum

Old man's beard stonecrop

You'll know this one by its fat little rosettes covered with stiff white hairs—earning its common name. Each leaf measures around ½ inch (1 cm) wide and over 1 inch (2.5 cm) long. When it blooms, the plant looks most like a sedum, sporting plenty of single white flowers accented with extra-large red anthers. Evergreen. Propagate by cuttings.

ZONES 10–11

PLANT SIZE 3–5 inches (7.5–13 cm) tall, 5–6+ inches (13–15+ cm) across.

SOIL very dry

LIGHT full sun

SIMILAR SPECIES AND CULTIVARS Sometimes sold incorrectly as *Sedum hintonii*, another hairy Mexican species; that species is smaller and rare in cultivation.

ORIGIN Mexico

LANDSCAPE AND DESIGN USES This is an absolutely dry situation plant, best in a container (few outdoor locations in the United States would suit it). Because it can rot easily, water very sparingly and do not water at all when the weather is cool. Due to its white appearance, provide some contrast, either a dark-leaved plant or a medium to dark mulch of some kind. Protect this plant from freezes.

Sedum moranense

Red stonecrop

A highly variable species, but that is part of its eye-catching beauty. There is a slight glaucous cast to the leaves, giving the plant multiple colors that contrast with its red stems. The entire plant can turn maroon to purple in winter. The habit is upright then sprawling. Thick but small, blunt-edged leaves that come to a point remind me of shark teeth in their shape; they spiral along the stems at almost right angles. Blooming occurs in late winter or early spring in small clusters at the tips; flowers are five-petaled and white with red anthers. This plant is vigorous, quite adaptable, and deserves more use. Evergreen. Easy to propagate by cuttings.

ZONES 6–9

PLANT SIZE 4–7 inches (10–18 cm) tall, 18 inches (45 cm) across.

SOIL dry to well-drained

LIGHT full sun

SIMILAR SPECIES AND CULTIVARS *Sedum moranense* subsp. *grandiflorum* has larger foliage and flowers twice as large as the species, although foliage of the subspecies is bright green and glossy, which makes me think the plant is a hybrid. *Sedum* 'Spiral Staircase' is larger still. Resembles *Sedum acre*, except the leaves are not as congested in the lower stems, or even bare.

ORIGIN Mexico

LANDSCAPE AND DESIGN USES It would blend nicely with the silver foliage of lavenders. The species is surprisingly hardy, considering it hails from Mexico.

Sedum morganianum

Burro's tail, donkey tail

This plant's trailing habit has made it probably the most common houseplant sedum since its introduction into cultivation in 1935. Thick, fingered foliage with a distinct curve toward the pointed tips is whorled around hanging stems that can eventually reach up to 3 feet (90 cm) long. The blue-green glaucous foliage is also distinct unless you're looking at some of its relatives. In winter, some of the lower foliage can easily fall off the stem. Flowers appear in late spring, only on stems that have reached a certain size, usually around 1 foot (30 cm) long. Dark red flower clusters hang away from the foliage at the tips and never fully open. Evergreen. Propagate by stem or leaf cuttings.

ZONES 10–11

PLANT SIZE 12–36+ inches (30–90 cm) tall, 12 inches (30 cm) across.

SOIL dry to well-drained

LIGHT full sun

SIMILAR SPECIES AND CULTIVARS *Sedum treleasei* has thicker foliage. *Sedum burrito* is a closely related, slower-growing species with 1-inch (2.5-cm) leaves that are thicker, closer, and not pointed. ×*Sedeveria* 'Harry Butterfield', a hybrid with *Echeveria derenbergii*, is known as super burro's tail. There's also a hybrid with *Sedum treleasii* that has the common name of giant burro's tail, sometimes sold as *Sedum* 'E.O. Orpet'.

ORIGIN Mexico

LANDSCAPE AND DESIGN USES This plant is extremely drought-tolerant and can be long-lived. Surprisingly, it comes from humid regions where it grows in the shade. I like it with other strongly whorled plants like the other donkey tail (*Euphorbia myrsinites*).

Sedum multiceps

Miniature Joshua tree stonecrop

For its size, this one has a lot of character. It is strongly branched and matures to look like a little Joshua tree. The entire plant has a shiny yet fuzzy look due to papillose foliage, which means it has miniature bumps on the leaf surfaces. Individual leaves are paler at the base, turning light green where they meet the stem. Lower foliage dries brown but holds onto the stem below the growing tips. Flowers are a pleasing creamy yellow with up to seven petals. Evergreen. Propagate by cuttings in fall or winter.

ZONES 9–11
PLANT SIZE 6 inches (15 cm) tall, 2–3 inches (5–7.5 cm) across.
SOIL well-drained, average
LIGHT full sun
SIMILAR SPECIES AND CULTIVARS This species is similar botanically to *Sedum acre* and *S. sexangulare*, but not in form or habit.
ORIGIN Algeria, but is found throughout the Mediterranean.
LANDSCAPE AND DESIGN USES Its upright habit makes it a nice container plant, and compared to many alpines, this one will stand out. Due to its size, it is best planted with other small plants including *Sedum japonicum* 'Tokyo Sun' and *S. japonicum* var. *pumilum*.

Sedum nokoense

SYNONYM *Sedum taiwanianum*

Relatively new to the United States, but sure to become popular. The dark green, glossy leaves and ground-hugging habit make this a nice groundcover for small areas. Relatively small yet thick spoon-shaped foliage measures only around ½ inch (1 cm) long and not quite as wide; close inspection reveals that each leaf has a distinct crease down its center. Reddish stems grow 4 inches (10 cm) long with congested whorled foliage at the tips. The plant blooms in loose yellow inflorescences in late summer into fall and winter. Individual flowers have between four and six thin, linear petals. Carpels are green turning red or orange as the petals fade. Evergreen. Propagate by division.

ZONES 8–11, maybe colder
PLANT SIZE 2–4 inches (5–10 cm) tall, 3–4+ inches (7.5–10+ cm) across.
SOIL rocky, dry to well-drained, average
LIGHT full sun
SIMILAR SPECIES AND CULTIVARS *Sedum nokoense* 'Cidense' is a compact plant. Similar to *Sedum makinoi.*
ORIGIN the mountains of southern Honshu, Japan; Taiwan; Korea.
LANDSCAPE AND DESIGN USES I first saw this species planted in a trough with evergreens— perfect. Just remember, due to its dark green foliage, it can get lost without contrasting gravel mulch or lightly colored plants next to it. It should do well in hot, humid climates.

Sedum oreganum ▾

Oregon stonecrop

The glossiness of the foliage and its contrasting colors make quite an attractive combination. The shape of the leaves reminds me of a club; older leaves are tipped red. Habit is creeping, and the spiraling foliage ends in rosettes. The yellow flowers fit nicely with the green and red foliage. Evergreen. Propagate from cuttings or seed, in fall or spring.

ZONES 5–9
PLANT SIZE 3–4 inches (7.5–10 cm) tall, 6+ inches (15+ cm) across.
SOIL moist, cool
LIGHT full sun to part shade
SIMILAR SPECIES AND CULTIVARS *Sedum oreganum* subsp. *tenue* is a choice smaller form that stays red much of the time. *Sedum divergens* is similar. *Sedum* 'Helen Payne' looks like a cross with *S. divergens* or *S. spathulifolium*.
ORIGIN Pacific Northwest (United States and Canada).
LANDSCAPE AND DESIGN USES This plant can decline in the heat of summer if it doesn't get some shade and enough moisture. It is an ideal candidate for use on a greenroof in the Pacific Northwest.

Sedum oaxacanum ▴

A good-looking trailer, suitable for containers and warmer climates. Fleshy red stems are lined with green foliage; the tips turn upright and the foliage is more congested and more strongly glaucous. Individual leaves are flat-topped but round-bottomed and about twice as long as wide. In summer, perky yellow flowers are borne singly or in small clusters. Evergreen. Propagate by cuttings.

ZONES 9–11
PLANT SIZE 3 inches (7.5 cm) tall, 6+ inches (15+ cm) across.
SOIL dry to well-drained, average
LIGHT full sun
SIMILAR SPECIES AND CULTIVARS *Sedum australe, S. wrightii.*
ORIGIN Mexico
LANDSCAPE AND DESIGN USES Its relaxed stems make it ideal for overhanging an edge or a container. It has an open habit, so it needs to be planted with something to complement that, such as *Phedimus spurius*.

Sedum oregonense

Cream stonecrop
SYNONYM *Sedum rubroglaucum, S. watsoni*

The most unique feature of this plant is its blue-green leaves that are notched at the tips. The leaves are also glaucous, opposite, and alternate from the previous pair of leaves by ninety degrees (decussate). Thick stems grow horizontally and tend to have some old foliage attached. Flower stalks develop in early spring (note that the foliage on these is not notched). The flowers are quite unique-looking: each has five hooded yellow petals with green bases and green carpels, and the stamens are yellow at the tips but green at the base. Look closely to detect another very distinguishing characteristic—the sepals are fused at the base. Evergreen. Propagate by division.

ZONES 5–9
PLANT SIZE 3+ inches (7.5 cm) tall, 5+ inches (13+ cm) across.
SOIL rocky, gravelly, dry to well-drained
LIGHT full sun
SIMILAR SPECIES AND CULTIVARS *Sedum laxum* is the only other North American species with notched foliage; however, its flowers are pink or white.
ORIGIN Cascade Mountains in Oregon.
LANDSCAPE AND DESIGN USES In nature, this species has little competition and in garden situations, it has good vigor. Pair it with a red-leaved *Phedimus spurius*, or mix it with other green Pacific Northwest natives such as *Sedum divergens* and *S. oreganum*.

Sedum pachyphyllum
Tree sedum, silver jellybeans

A bold look! Features thick, fingered foliage typically around ¼ inch (6 mm) thick. The tips are wider, arched up from the stem, and some are red. Color varies from blue-green to light green with a glaucous cast. Long stems display the yellow flowers in flat clusters; sepals are thicker at the tips and very uneven. Evergreen. Easily propagated from both leaf and stem cuttings.

ZONES 10–11

PLANT SIZE 20 inches (50 cm) tall, 6 inches (15 cm) across.

SOIL dry to well-drained, average

LIGHT full sun

SIMILAR SPECIES AND CULTIVARS *Sedum* 'Ron Evans', a hybrid with *S. pachyphyllum* as one parent, looks thicker and more congested. ×*Sedeveria hummellii*, a hybrid between *S. pachyphllum* and *Echeveria derenbergii*, has larger yellow flowers that fade to red.

ORIGIN Mexico

LANDSCAPE AND DESIGN USES After burro's tail, this is one of the most common houseplant sedums; ease of care makes it a staple in succulent collections. It will tolerate a frost when the soil is dry. Stunning when paired with red forms such as *Sedum* ×*rubrotinctum* and *S. stahlii*.

Sedum palmeri

Palm stonecrop

An appealing plant. Woody-looking brown to gray stems hold rosettes of gray-green at their tips; certain forms are flushed with red. The variable leaves can be oval-shaped, not quite twice as long as wide, with some minutely serrated edges that come to a tip. Flowers begin in late winter and are borne laterally. The yellow to golden yellow clusters have pointed petals. Evergreen. Propagate by cuttings.

ZONES (8)9–11, maybe colder

PLANT SIZE 8–12 inches (20–30 cm) tall, 15+ inches (38+ cm) across.

SOIL dry to well-drained, average

LIGHT full sun

SIMILAR SPECIES AND CULTIVARS This species varies quite a bit. There are small forms only growing around 4 inches (10 cm) tall. Two good ones are *Sedum palmeri* subsp. *emarginatum* and *S. palmeri* subsp. *rubromarginatum*. *Sedum compressum* is a synonym of the small form. *Sedum obcordatum* has an opposite, criss-crossed (decussate) foliage that is more rounded.

ORIGIN Mexico and Texas.

LANDSCAPE AND DESIGN USES This plant has lots of character and deserves more use. It does well both outdoors and inside as a houseplant. Because it has upright stems, I recommend underplanting it with a low-grower like *Sedum diffusum*. Interestingly, this plant seems to have been embraced in Italy—many photographs I came across showed it in use there as a patio plant.

Sedum polytrichoides 'Chocolate Ball'
Chocolate Ball stonecrop

Fine foliage and dark green to chocolate to bronze hues that ebb and flow give this plant a lot of character. The brownish hues hold over the summer months; I was surprised to see this plant turn mostly dark green over the winter in a cool greenhouse. Leaves are linear and whorled, like those of many *Petrosedum rupestre* plants. Crowded crowns have upright then sprawling stems. Characteristic of this plant, shaggy old foliage holds on at the base of the stems. The shy flowers are light yellow with short petals. Evergreen. Easy from division in spring or fall.

ZONES (6)7–9

PLANT SIZE 5 inches (13 cm) tall, up to 3+ inches (7.5+ cm) across.

SOIL moist, well-drained, average

LIGHT full sun

SIMILAR SPECIES AND CULTIVARS Often sold as *Sedum hakonense*, but this is the wrong name for this plant. At times, it can resemble *Petrosedum rupestre*; maybe that is why some think it is hardier than it actually is.

ORIGIN Kyushu Island, Japan; Korea.

LANDSCAPE AND DESIGN USES The dark foliage makes this plant an ideal contrast for many other sedums. It would look terrific with *Sedum mexicanum* 'Lemon Ball' or *S. japonicum* 'Tokyo Sun'. Although it is not cold-hardy in Zone 5, gardeners there could bring a small piece in for the winter and start it again in spring.

Sedum praealtum
Beavertail stonecrop, tree sedum

Another fine Mexican tree sedum. The best way to tell it apart from its close relative *Sedum dendroideum* is by its leaves. Leaves of *S. dendroideum* have window-like glands along the leaf margin, while leaves of *S. praealtum* are light green, long, flat, and have smooth edges. They attach to the stem at more or less right angles but arch up towards the tip. Flowers are carried in large clusters on long pedicels; they are yellow with thin petals. Evergreen. Propagate by cuttings.

ZONES (9)10–11
PLANT SIZE up to 36 inches (90 cm) tall, 12–15 inches (30–38 cm) across.
SOIL dry to well-drained
LIGHT full sun
SIMILAR SPECIES AND CULTIVARS *Sedum praealtum* subsp. *parvifolium* has leaves that are smaller and more rounded, more like the shape found in *S. dendroideum*.
ORIGIN Mexico
LANDSCAPE AND DESIGN USES This semi-hardy tree species makes a long-lived, trouble-free container plant that resembles a miniature tree. In warm climates, it can be grown outdoors. With its stemmed habit, it allows spaces for planting other succulents around its base. Nice companions include aeoniums. crassulas, graptopetalums, and pachyphytums. This tree sedum can be knocked back to the ground after a freeze, so when planting it, keep in mind that winter location strongly determines the plant's ultimate height.

Sedum pulchellum
Widow's cross, sea star

This is a favorite of mine. It has a couple of leaf shapes due to its biennial nature and acts like a winter annual. It flowers in summer, with unique arching pink sprays that look like jester hats. Then it goes to seed and the plants slowly dry to cinnamon brown. By late summer to early fall, seedlings begin to appear. These grow into rosettes with spoon-shaped leaves on short stems (petioles) loosely whorled around the tips. By spring, the rosettes begin to expand, adding linear, whorled, bright green foliage above some usually red old foliage below. The new plants bloom in early summer and the process starts over. Once flower stems are erect, you'll notice strongly clasping (amplexicaule) leaves. Evergreen. Propagate by seed.

ZONES (3)4–9
PLANT SIZE 6 inches (15 cm) tall, up to 8–10 inches (20–25 cm) across.
SOIL moist, average and, in its native habitat, thin

LIGHT full sun to part shade
SIMILAR SPECIES AND CULTIVARS The species is sometimes sold as *S. pulchellum* 'Sea Star'.
ORIGIN Eastern United States west to the Mississippi basin.
LANDSCAPE AND DESIGN USES You will like the way the plant moves around and fills in holes by reseeding throughout the garden; once you have it you should always have it. (If plants appear where they are unwanted, they are easily rouged.) The facts that this is a United States native and that it can take some shade are other reasons I recommend it. It's irresistible in combination with *Phedimus kamtschaticus* var. *floriferus* 'Weihenstephaner Gold'. I enjoy it in an irrigated, full-sun garden next to *P. obtusifolius* var. *listoniae*. Both plants have pink flowers in summer, plus nicely contrasting foliage. In spring, this plant can appear to be a *Petrosedum rupestre*, acting like a chameleon at times.

Sedum ×rubrotinctum

Jellybean plant, Christmas cheer, pork and beans

A colorful and dependable plant with chubby leaves usually around 1 inch (2.5 cm) long and approximately ¼ inch (6 mm) wide, tapering at the ends. The red color starts at the tips. In summer, the plants are mostly red, but they tend to green up in the winter. The foliage spirals around the stem, becoming more congested at the top of the plant, while some of the older foliage drops off to leave most of the lower stems bare. Not really grown for its flowers, as they can be shy to appear (but if they do, congested clusters of yellow, orange, and red buds open to yellow flowers with strongly pointed petals). A tender perennial that can survive an occasional freeze. Evergreen. Propagate by cuttings. Received the Award of Garden Merit (AGM) from Britain's Royal Horticultural Society in 2012.

ZONES (9)10–11
PLANT SIZE 4+ inches (10+ cm) tall, 6–10+ inches (15–25+ cm) across.
SOIL dry to well-drained, average
LIGHT full sun
SIMILAR SPECIES AND CULTIVARS This plant is a cross, probably between *Sedum pachyphyllum*, which has silver foliage, and *S. stahlii*, which has small, jellybean-shaped foliage of brick red. Pink jellybeans, *S. ×rubrotinctum* 'Aurora' (synonym 'Vera Higgins') is a sport with glaucous gray foliage that changes to pink in sun; it deserves more use. *Sedum guatemalense* has pink flowers and flattened gray-green foliage. *Sedum* 'Joyce Tulloch' has a more upright, open habit. *Sedum* 'Little Gem' is also similar.
ORIGIN garden origin
LANDSCAPE AND DESIGN USES Commonly enjoyed as a houseplant. Inside or out, it makes an easy-care container plant and a great companion to other succulents.

Sedum sarmentosum
Stringy stonecrop, whorled stonecrop

The foliage is memorable because it is so light green as to almost be yellow at times. Individual leaves are lance-shaped (lanceolate), carried in sets of three at well-spaced intervals along peachy pink, speckled stems. Flowers are shy but distinct, looking semidouble due to (usually) six sepals of a similar color and size to the yellow petals, centered with yellow ovaries or carpels and dark anthers. In winter, most of the foliage falls off except at the tips, leaving the bare stems. Some of the tips will die back, but these vigorous plants will regrow rapidly in spring, regaining lost ground. This plant is mostly deciduous except for small, ground-hugging green rosettes in winter. Propagate by cuttings.

ZONES 4–9
PLANT SIZE 3–4 inches (7.5–10 cm) tall, 24 inches (60 cm) across.
SOIL well-drained to moist
LIGHT full sun to part shade
SIMILAR SPECIES AND CULTIVARS Sometimes nurseries offer this plant as "gold moss stonecrop," which is also a common name for *Sedum acre* and can create confusion and misnaming. *Sedum lineare* is similar. The stems of *S. sarmentosum* remind me of *S. mexicanum*, which might be a parent. This plant is sterile and could actually be a hybrid.
ORIGIN China
LANDSCAPE AND DESIGN USES In the right spot this plant is hard to beat, if you like the light color and the way it wraps around stones and other plants. Be forewarned, however, that it can take over smaller, weak plants in the shade (I've never seen it be as aggressive in a full-sun situation). The long stems make this an ideal choice in places where they can hang down, namely, hanging baskets, windowboxes, large containers, and on shaded slopes.

Sedum sexangulare

Tasteless stonecrop

As the scientific name suggests, the spiraling, fingered foliage usually has six sides, though it can vary from five to seven. The more shade, the greener the plant will remain. In full sun, in winter, foliage can become copper colored, but it greens up with new growth in spring. Older, lower leaves will dry out and remain on the lower parts of the stems. The plant blooms in early summer; flowers are yellow. Evergreen. Easy to propagate by division, cuttings, or seed.

ZONES (3)4–9

PLANT SIZE 4 inches (10 cm) tall, 15 inches (38 cm) across.

SOIL well-drained, average

LIGHT full sun to part shade

SIMILAR SPECIES AND CULTIVARS This species can be easily confused with *Sedum acre*. The main difference is the shape of the foliage; *S. sexangulare* leaves taper at both ends (are "terete"), while those of *S. acre* typically have a broad base. (Another way to tell is to taste it—*S. acre* will be bitter.) *Sedum acre* 'Aureum' is similar. *Sedum sexangulare* 'Golddigger' is a lighter green, chartreuse form that yellows before flowering; it was introduced by the author in 2011. *Sedum sexangulare* 'Red Hill' turns orange in winter; 'Utah' is dark green; 'Weisse Tatra' is a more compact form. *Sedum sexangulare* subsp. *montenegrinum* and *S. tschenokolevii* are both almost identical.

ORIGIN Central Europe

LANDSCAPE AND DESIGN USES This is one of the most adaptable species. It is a staple sedum for greenroofs.

Sedum spathulifolium subsp. *pruinosum* 'Cape Blanco'

Pacific stonecrop, broadleaf stonecrop
SYNONYM *Sedum spathulifolium* 'Cape Blanco'

Despite the debates or mix-ups about the correct name, this is a popular and widely sold plant, particularly in the Pacific Northwest. There is also a handful of closely related plants (see below), further complicating matters. Compared to the species, 'Cape Blanco' is a more robust grower. In any event, this ground-hugging plant is prized for its silvery, or silver-and-green coloration. Spoon-shaped (spathulate) leaves whorl around the tips, forming silver rosettes on creeping stems. In early summer, short, extra-thick stems support clusters of yellow flowers that are attractive to butterflies (in fact, it's a host plant for two species). These appear to be self-sterile, suggesting that the plant could be a hybrid. Evergreen. Propagate by cuttings.

ZONES 5–9
PLANT SIZE 4 inches (10 cm) tall, 8–12 inches (20–30 cm) across.
SOIL open, rocky, dry to well-drained
LIGHT full sun to part shade
SIMILAR SPECIES AND CULTIVARS There are some related hybrids made by Helen Payne that have similarities and are worth searching for. *Sedum* 'Moonglow' is probably a cross between *S. laxum* subsp. *heckneri* and *S. spathulifolium* subsp. *spathulifolium* (it appears to have characteristics of both plants). *Sedum* 'Silver Moon' is thought to be a backcross, meaning the same parents were used but the seed parent and pollen parent were reversed. *Sedum* 'Harvest Moon' is a cross between *S.* 'Silver Moon' and *S. spathulifolium* 'Carnea'. A variety called 'White Chalk' (so named by Wayne Fagerlund of Evergreen Valley Nursery in Washington State) is probably just the species that has been propagated vegetatively.
ORIGIN garden origin
LANDSCAPE AND DESIGN USES This makes a nice groundcover, thriving especially in the Pacific Northwest. With some shade, it should work on a greenroof there, too. Combining its silver foliage with a red-leaved selection such as *Phedimus spurius* 'Fuldaglut' (fireglow) would be gorgeous. Although hardy, the plants do not overwinter reliably in the regions with wet and freezing winters.

Sedum spathulifolium subsp. *purpureum*
Pacific stonecrop

The red and silver combination of this subspecies is dazzling, reminiscent of jewels. Spoon-shaped leaves congregate at the tips to form cute red and chalky silver rosettes. Plants spread by thin aboveground shoots. The yellow flowers can vary but bloom in clusters on short stems only around 4 inches (10 cm) tall. Evergreen. Propagate by cuttings.

ZONES 5–9

PLANT SIZE 2–4 inches (5–10 cm) tall, 8–10+ inches (20–25+ cm) across.

SOIL rocky, dry to well-drained

LIGHT full sun to light shade

SIMILAR SPECIES AND CULTIVARS This plant was incorrectly given a cultivar name of 'Red Chalk' and is still sometimes sold by that name. *Sedum spathulifolium* 'Carneum' is one of the most common red-leaved forms. *Sedum spathulifolium* 'Aureum' has buttery yellow leaves that develop red highlights in more sun while retaining the white coating (pruinose) effect, but it is not as hardy. *Sedum spathulifolium* 'Rogue River' has thick blue foliage. *Sedum spathulifolum* subsp. *purdyi* is a smaller green form but retains some of the white coating on its leaves. *Sedum yosemitense* used to be considered a subspecies and is supposedly more tender, even though it is from the Sierra Nevada range; it has red highlights with smooth-textured green foliage, and is one of the few nonpruinose forms. *Sedum yosemitense* 'Red Raver' has brilliant red stems and olive green leaves.

ORIGIN Pacific Northwest (United States and Canada).

LANDSCAPE AND DESIGN USES This plant is commonly used at the edge of low walls, in the company of other plants that hang over, such as sunrose or rock rose (*Helianthemum* spp.).

Sedum stahlii
Coral beads

A stunning plant. Much of the time it is a dark red color, but it can brighten to coral, or green up, depending on the time of year. Beadlike foliage, which grows in opposite pairs, has a very fine peach fuzz appearance (pubescent). Stems start upright but quickly relax and eventually sprawl, forming a loose mat, or hang down. Light yellow flowers with five petals each really stand out against the foliage. Evergreen. Propagate by stem or leaf cuttings.

ZONES 10–11
PLANT SIZE 6–8 inches (15–20 cm) tall, 12+ inches (30+ cm), flower stems up to 36 inches (91 cm) at maturity.
SOIL well-drained to dry
LIGHT full sun
SIMILAR SPECIES AND CULTIVARS *Sedum stahlii* 'Variegata' is rare.
ORIGIN Mexico
LANDSCAPE AND DESIGN USES This species makes a great companion to both yellow-leaved and bright green creeping sedums like *Petrosedum rupestre* 'Angelina'. It would also serve well as a basal cover plant to other super-succulent species with upright or rosette habits such as *Sedum clavatum*, *S. lucidum*, *S. palmeri*, and *S. treleasei*.

Sedum stenopetalum 'Douglasii'

Wormleaf stonecrop
SYNONYM *Amerosedum stenopetalum* 'Douglasii'

This cultivar was chosen due to its stability and greater reliability compared to the species. It is a mat-former with an attractive combination of red and green foliage. At maturity, it may take on a pyramidal, upright shape. Leaves are narrow and held in whorls around the stem in a spiral pattern; they are flat above and rounded below. In summer, the top foliage takes on orange-red coloration, while older foliage below dries to a cinnamon color but remains on the stem. The cultivar is less likely to flower than the species, but if it does, all flower parts are yellow and the petals are strongly ridged. Instead, this plant prefers to spread via adventitious buds that form along the stems. These fall off easily and blow around until they find a good spot to root or grow. Evergreen. Propagate by division.

ZONES (3)4–9
PLANT SIZE 6+ inches (15+ cm) tall, 5–6+ inches (13–15+ cm) across.
SOIL rocky, dry to well-drained
LIGHT full sun
SIMILAR SPECIES AND CULTIVARS It can look like *Petrosedum rupestre* at times, especially as a seedling, but its leaves are more linear. *Sedum lanceolatum* var. *rupicolum* is another North American native that flowers earlier and has flatter foliage.
ORIGIN United States and Canada.
LANDSCAPE AND DESIGN USES An excellent choice for a rock garden or a greenroof, because it can spread and fill in gaps, thanks to the little plantlets it forms on the flower stems. Does especially well in the Pacific Northwest.

Sedum suaveolens

Sweet-smelling stonecrop
SYNONYM *Graptopetalum suaveolens*

The oddest, most surprising fact about this plant is something you can't see: it has the largest number of chromosomes of any living thing on earth. In its rosette stage (that is, when it is not in bloom), it strongly resembles *Echeveria elegans*. The leaves are blue to pink-purple, but sometimes so glaucous as to appear almost white. Stolons emerge from among leaves with new rosettes that root down next to the mother plant, forming a small colony. Short flower stems also emerge from the foliage but hardly reach the edge of the plant; these bear sweet white flower clusters that bloom in summer. Upright petals are curved back at the tips. Evergreen. Propagate by removing new shoots from the mother plant and planting, or raise from seed.

ZONES 10–11
PLANT SIZE 3+ inches (7.5 cm) tall, 6 inches (15 cm) across, flower stems 6+ inches (15+ cm) tall.
SOIL dry to well-drained
LIGHT morning sun to bright shade
SIMILAR SPECIES AND CULTIVARS *Sedum craigii* also has silver rosettes, but is smaller. *Echeveria elegans* might look like it in foliage, but not in flower.
ORIGIN Mexico, in the western Sierra Madre.
LANDSCAPE AND DESIGN USES Plant these strong silver rosettes with other similar plants like *Echeveria*.

Sedum ternatum

Woodland stonecrop, three-leaved stonecrop, whorled stonecrop

This stonecrop is easy to recognize by its prolific spring flower show. Branched and arching open cymes carry four-petaled white flowers; petals are thin compared to many, sepals are large, and dark red stamens provide contrast. Foliage is spoon-shaped in groups of three around the stems, congesting into rosettes at the tip. The foliage is generally green. Sometimes the stems can get an orange-red tinge that bleeds into some of the older leaves. Evergreen. Propagate by cuttings, division, or seed.

ZONES 4–8
PLANT SIZE 3–6 inches (7.5–15 cm) tall, 6–9+ inches (15–23+ cm) across.
SOIL well-drained, average to moist
LIGHT full sun to part shade
SIMILAR SPECIES AND CULTIVARS *Sedum ternatum* 'Larinem Park' is a larger and more robust form. White forms of *Phedimus spurius* are similar to this species, but summer-blooming.
ORIGIN Eastern United States.
LANDSCAPE AND DESIGN USES It makes a nice groundcover especially in a moist, part-shade environment. It doesn't appreciate full-sun exposure and will deteriorate in the heat of summer. Southern Illinois University tested it on a shaded greenroof and found it to be very reliable. One of my favorite combinations was spontaneous—*Viola labradorica*, which has dark green foliage and purple spring flowers, seeded into my *Sedum ternatum*, which bloomed white at the same time. A simple pairing, but it makes an unforgettable blend of texture and color.

Sedum tetractinum 'Coral Reef'
Chinese stonecrop

Aptly named! Dark green leaves line long trailing stems of pink to coral. They take on some coral highlights in full sun, and in fall, the whole plant turns coral. The foliage is nearly round and is opposite. The flowers can be sporadic and are not very showy; they appear in summer and are yellow, with four or five petals. Evergreen. Propagate from cuttings.

ZONES 6–8
PLANT SIZE 3–4 inches (7.5–10 cm) tall, 12+ inches (30+ cm) across.
SOIL dry to well-drained, average
LIGHT part shade, but tolerates full sun
SIMILAR SPECIES AND CULTIVARS The foliage of *Sedum tetractinum* remains green in winter. *Sedum makinoi* has a similar appearance. There is a small form called 'Little China' from Paul Little of Little Hill Nursery in Tennessee.
ORIGIN China
LANDSCAPE AND DESIGN USES A good groundcover, thanks to its fast growth rate. The trailing stems also make it a choice hanging plant. In milder climates such as the Pacific Northwest, it can be used on greenroofs. Plant it with the species, whose leaves remain green in winter and thus contrast nicely with this selection's reddish hues.

Sedum treleasei
Silver sedum

This one has a very chubby look. Extra-thick, almost wedge-shaped leaves are glaucous blue-green. The stems eventually grow upright. Yellow flowers appear in round clusters in early spring; they have six petals and uneven green sepals. Evergreen. Propagate by division or seed.

ZONES 10–11
PLANT SIZE 6 inches (15 cm) tall, 12 inches (30 cm) across, flower stems 20+ inches (50+ cm) tall.
SOIL dry
LIGHT full sun, except in the heat of summer in hot, dry climates
SIMILAR SPECIES AND CULTIVARS This plant is sometimes offered as *Sedum* 'Lime Gum Drops'. Two similar plants are *S. commixtum* and *S. macdougallii*. *Sedum* 'Haren' is a hybrid of *S. treleasei* and *S. pachyphyllum*.
ORIGIN Mexico
LANDSCAPE AND DESIGN USES This is an excellent container plant, but because of its upright habit, you should use a substantial pot so it doesn't tip over as it matures. The thick foliage goes great with other thick succulent sedums of contrasting colors; a couple of good choices would be *Sedum adolphii* and *S. lucidum*.

Sedum wrightii

Wright's stonecrop

The flowers are especially pretty and nicely fragrant, too. They're white to pale pink, with upright petals that spread at the tips to resemble a chalice. The stamens and carpels are upright as well. These are borne on short, branching stems above stalkless rosettes of thick leaves. Foliage color is light green with a hint of silver. The roots creep along to form a tight colony. The flower stems form in spring, lengthen over the summer, bud in fall, and bloom in winter. Evergreen. Propagate by division or seed.

ZONES 10–11
PLANT SIZE 4 inches (10 cm) tall and wide.
SOIL rocky

LIGHT part sun to light shade
SIMILAR SPECIES AND CULTIVARS *Sedum australe* is smaller and finer; *S. multiflorum* is larger, growing to 12 inches (30 cm) tall, with more linear basal foliage, brittle red stems, and lots of starry white flowers with some red in them.
ORIGIN Mexico, into Texas and New Mexico.
LANDSCAPE AND DESIGN USES This plant naturally grows under trees or on north slopes in rocky soil, and it will want similar conditions in a garden. It can rot if the soil is too wet. Try it under the tree sedums— *Sedum pachyphyllum* or *S. palmeri*.

Sinocrassula yunnanense

Chinese crassula
SYNONYM *Sedum indicum* var. *yunnanensis*

Tiny and intriguing. Tight rosettes only measure an inch (2.5 cm) or so wide but, in time, cluster in groups 3–4 inches (7.5–10 cm) wide. Thin, pointed foliage ranges from dark green to brown, and usually is speckled with dark spots. Flowers appear in summer; these are creamy white with only one row of stamens compared to the normal two rows in *Sedum*. The stamens alternate between the petals. Evergreen. Propagate by division or seed.

ZONES (6)7–9
PLANT SIZE 1 inch (2.5 cm) tall, 2–4 inches (5–10 cm) across, flower stems 3–4 inches (7.5–10 cm) tall.
SOIL well-drained to dry
LIGHT full sun

SIMILAR SPECIES AND CULTIVARS *Sinocrassula indica* has wider foliage. *Prometheum pilosum* also has small green rosettes of thin, hairy foliage but tiny, pink, tubular flowers.
ORIGIN Nepal, western China at high altitudes.
LANDSCAPE AND DESIGN USES This plant is mostly used in small containers and rock gardens. It is ideal in a crevice garden. Plant it in a spot that can be protected from winter wetness. Light-colored gravel mulch is helpful to keep the lower foliage dry and to provides contrast to the dark foliage (which can get lost in an organic soil). Perfect with other small rosette-forming plants like jovibarbas, rosularias, and sempervivums.

GROWING
AND
PROPAGATING

F

For the most part, sedums are easy to plant, easy to care for, and easy to propagate. As garden citizens, they are attractive and resilient. That said, there are still some basics you should know to bring out the best in them.

Hardiness

The very first thing to consider before you choose and install any plant is, will it survive in your area?

A sedum's ability to withstand low winter temperatures or even freezing or snowy conditions depends on its origins. Those that hail from the Mexican desert, and their derivatives and offspring, of course, are not going to make it up north—although if that's where you live, you can always try less-hardy plants in pots and bring them indoors when winter approaches.

Note that the plant descriptions in this book include the known USDA Hardiness Zone information. This information should also be supplied by local and mail-order nurseries; when in doubt, ask. Of course, you can experiment with the hardiness of some plants, perhaps one zone to the north or south, say, if you have warmer microclimates on your property, or are willing to coddle a borderline plant. Pay attention to where a plant originated, for this will help you decide if the plant will grow in your area, or if you have to bring it inside for the winter or protect it from the heat of summer. Find one or several that attract you and give them a try.

Generally, the border sedums (*Hylotelephium* species and cultivars) behave like many other popular garden perennials and perform well in Zones 4 or 5 to 9. Winter mulch is advisable in Zones 4 and 5.

The dried stems of *Hylotel-ephium* 'Herbstfreude' add textural interest in the winter garden.

Semihardy Stonecrops for Zone 7 and South

Sedum clavatum
Sedum diffusum
Sedum makinoi cultivars
Sedum mexicanum cultivars
Sedum moranense
Sedum ×rubrotinctum
Sedum tetractinum cultivars

Hardy Stonecrops for Zone 6 and North

Petrosedum rupestre cultivars
Phedimus ellacombeanus cultivars
Phedimus hybridus 'Immergrünchen'
Phedimus kamtschaticus cultivars
Phedimus spurius cultivars
Rhodiola pachyclados
Sedum acre cultivars
Sedum album cultivars
Sedum lydium
Sedum oreganum
Sedum sexangulare cultivars

Check my landscaping recommendations and create scenes and combinations that look great. Finally, in the back of the book, you'll find useful resources, including sources for these plants and places to go for more details. If you are seeking a specific plant or piece of information, please consult the index.

Site and Sun

From the small, groundcovering stonecrops types to the big, upright border types, most sedums relish a spot in full sun. Remember, they are succulents; sunshine brings them the warmth they like and, thanks to their thick leaves, does not stress them. So

choose a site for them anywhere they are exposed to at least six full hours of sunlight a day or more.

There are exceptions. Some sedums do appreciate a little sheltering shade from blazing noontime summer sun, especially if they are not from a line of desert natives. If they are not protected, they dry out, slow or halt growth, or develop burnt tips or dry leaf edges. Others tolerate part shade, which simply means that they'll enjoy some hours in the sun and still be fine if lengthening shadows from your house or trees fall over them at other times—flexibility or adaptability many gardeners will appreciate.

The other requirement is a site with good drainage. If you already have sloped ground or rocky soil, you should have an ideal situation. Otherwise, create a raised area by mounding up soil, taking care to make it big and broad enough so that it doesn't look out of place or unnatural. Or grow sedums in a raised bed, no slope or change in elevation required. In one corner of our nursery, we did both: we grow cuttings in a raised bed with 6–8 inches (15–20 cm) of height gain from the path to the top of the bed.

Soil

Most sedums like to grow in well-drained to dry ground. Some hail from rocky slopes or desert landscapes and are satisfied with similar settings in your garden. Lean or less-fertile soil is usually fine. Their main enemy is soil moisture, which can lead to root rot or nurture plant diseases.

You might think a spot is fine when you first bring home some sedums only to lose them over the winter, when moisture can bog down in the soil, especially in ditches or low-lying parts of your landscape. Slopes or slight inclines are ideal because excess water can drain away easily. If you have concerns about the appropriateness of the soil in your chosen spot, scoop up a handful and have a look. If it's mucky or heavy, lighten it by digging in some sand, perlite, gravel chips, other material recommended by your local garden center, or a combination. Bear in mind that overly rich soil can lead to lanky, floppy growth and sometimes more foliage than flowers.

Whether you grow sedums in your garden ground or in a raised bed, make sure the soil itself is well-drained. Improve drainage by adding organic matter. A top-dressing of 1–3 inches (2.5–7.5 cm) of compost once or twice a year should be sufficient. As the soil improves, you can reduce such additions to only once a year. As a result of increased organic matter, your soil will have increased aeration. This also increases its water and nutrient capacity.

Soil pH can be an issue. A pH of 6.5 or higher is ideal for most sedums. In fact, in nature, many are native on alkaline soil (pH of 7 and higher). Should you add sand, gravel, or rocks to the area, take care that your choices are not too acidic. Otherwise, you'll have to add lime to neutralize their effect and raise the level—extra work best avoided.

Sedums for Shade

Hylotelephium populifolium
Hylotelephium sieboldii
Petrosedum forsterianum subsp. *elegans*
Phedimus ellacombeanus
Sedum glaucophyllum
Sedum makinoi cultivars
Sedum oreganum
Sedum pulchellum
Sedum sarmentosum
Sedum sexangulare
Sedum ternatum
Sedum tetractinum cultivars

The needled foliage of a sedum pairs well with the paddlelike leaves of prickly pear cactus (*Opuntia humifusa*). Both plants thrive with little water.

Planting

Sedums may be planted in spring, summer, or fall, though it's never wise to subject them to undue stress by planting on a hot summer day. Prepare the area ahead of time, removing all weeds, large roots, and other obstructions.

Ideally plants to be installed will be growing in similar-size containers. When you're ready, remove the sedums from their pots and gently loosen their root balls if dense and tight. Spacing is a matter of what you have in mind and which other plants are meant to be companions. Most sedums are not rampant, aggressive growers, so give them elbow room based on their predicted mature size. Site them a safe distance from bulkier, fast-growing plants like other perennials or grasses.

If you want a groundcovering stonecrop to fill in an area more quickly, buy more plants and plant them closer together but not touching. Many of the bigger border sedums, for instance *Hylotelephium* 'Matrona', should be planted no closer than 15 inches (38 cm) apart to allow for future expansion. Crowding such plants can lead to poor performance or distorted appearance. Place your plants in the ground at the same depth they were growing in the nursery pot. For the first couple of weeks, water newly planted sedums every day to help them get established.

Moisture, Mulch, and Fertilizer

Sedums are, for the most part, drought-tolerant, but they can't live without water. Moisture should move on through the soil after plant roots take what they need. Poor drainage is fatal to sedums. Overwatering can also be fatal—in fact, it is the easiest way to kill a sedum. Growing them in conjunction with other succulents and drought-tolerant plants is your best bet to assure your plants remain healthy and are not getting overwatered. Note that plants with gray to silver foliage tend to be more drought-tolerant.

In the garden, sedums will require much less water. At the nursery, we use a daily automatic watering system when the weather is hot, but at a far lower rate compared to other plants being grown in the same soil or conditions.

As a rule, sedums should not be watered if the soil is still moist from the previous watering. Sometimes you can tell it is time by looking at the soil; you will see a difference in color (wet soil is darker). You can also feel the top of the soil.

After your sedums are planted, there are a couple of things you can do—in addition to the minimal watering described above—to help them prosper. An excellent way to make a site welcoming to sedums is to add a top-dressing of rocky "mulch," an inch (2.5 cm) or more thick. Depending on availability, this could be quartzite or small, pea-size gravel around ⅜ inch (9 mm) in diameter. Note that angular gravel is more effective than washed or round material. Depending on the look you prefer, this material can be worked into the top layer of the soil or used like mulch. Over time, it will naturally work its way down into the soil. In the meantime, it makes a display look neater and more attractive.

Traditional garden mulches can also be beneficial to sedums, but don't overdo it. Never crowd their crowns, that is, don't push the material up against the plants' crowns

A stunted plant due to mulching on top of the plant's crown.

or stalks, which can restrict water and oxygen, raise local humidity, and lead to rot. It can also inhibit growth. Straw, dried grass clippings, chopped-up dried autumn leaves, cocoa hulls—any of these may serve if applied sparingly.

Whatever the material, a moderate mulch layer of an inch (2.5 cm) or more, depending on the plant size and the setting, can help tame fluctuating soil moisture, keep soil cooler, and discourage encroaching weeds. A thicker layer in winter in colder climates will help prevent frost-heaving.

When it comes to sedums, keep fertilizing to a minimum. In fact, I would not recommend using fertilizer in the garden at all unless your soil is extra-lean. That said, there are a few times when you might feed sedums. At the nursery, we give developing seedlings a little boost with some slow-release fertilizer worked into their soil mix. In the garden, you might want to feed a new groundcover to encourage it along or encourage a specimen plant to perform at its best.

In such cases, feed lightly and only when the plants are actively growing. Your best bet is to use a water-soluble fertilizer in a weak solution (perhaps a 5–10–10 at half-strength or less) and deliver the dose when watering anyway. Never feed in the fall. Sedums, like all perennial plants that are slowing down and going dormant for the winter, need to be allowed to do so.

This carefully selected duo provides color even in a hot, dry part of the garden.

Seasonal Maintenance

For the most part, sedums are undemanding, low-care plants. Here are a few things you can, and periodically should, do for them.

IN SPRING

Carefully remove winter mulch from groundcover areas and from around any hylotelephiums or autumn types.

After danger of frost is past, put tender potted sedums back outside in sunny spots.

Plant new arrivals.

If you have hearty, mature sedums in the ground or in pots, you can divide them now. Trim off any ragged-looking topgrowth or damaged roots, handle pieces gently, and water in the transplants well until they get established.

Optional: pinch or cut back hylotelephiums or autumn types. In England, cutting back a plant halfway to the crown in May in conjunction with the annual Chelsea Flower Show is termed the "Chelsea Chop." This increases branching and thus keeps an entire plant shorter and fuller. The drawbacks are that flowering can be delayed, and flowerheads will be smaller.

IN SUMMER

Pull weeds that are moving into your sedum areas.

Groom away any browning, damaged, or straying growth.

Remove spent flowerheads from groundcovering types if you don't like their look or don't want seedlings.

If any groundcovering type begins to exceed its bounds, check its wandering ways with a sharp spade or by trimming.

IN FALL

For hylotelephiums or autumn types, deadheading is rarely necessary. Indeed, their flowerheads are at their best now and remain attractive into the winter months. Only cut some if you are using them for bouquets.

Attack annual weeds in and among your displays or even just nearby, before they can go to seed. Yank out entire plants or, at least, remove the flowers. Also keep an eagle eye out for, and promptly remove, any small winter weeds or lawn grass that may be encroaching.

Grown for its evergreen foliage, *Phedimus takesimensis* makes a nice specimen in a container.

A perfectly matched plant and pot.

In a gravel garden, this is the time to cut back all the sedums. (In mine, I spare the showiest plants, such as the ornamental grasses, to enjoy over the winter months.) This keeps organic matter from building up.

Cut back deciduous sedums and tidy up around them.

If you garden in a cold climate or have some plants that are borderline hardy, mulch.

Move tender potted sedum displays indoors for the winter.

IN WINTER

If your area gets snow, make sure whoever is clearing walkways, paths, or driveways near your sedums does not liberally sprinkle road salt. Runoff will result in harmed or dead plants next spring.

Keep an eye on your potted displays resting indoors; be vigilant for insect pests. Water sparingly as needed only. Do not fertilize.

Order new plants and view new ideas in catalogs, books, magazines, or the Internet.

Containers

Whether you are putting together a small trough garden or filling a strawberry pot, use an appropriate soil mix so the plants will prosper. I recommend that you buy a bagged mix specifically labeled for use with succulents and cacti. Otherwise, you can certainly get traditional potting soil and amend it with perlite, pumice, or sharp gravel to provide extra drainage. At our nursery, we use a bark-based potting soil with rice hulls as an added ingredient to provide better drainage. Alternatively, if you'd like to mix your own, try this recipe: one part horticultural grade sand, one part grit (gravel bits, pumice, or perlite), and one part compost or standard potting soil.

Sedums in containers do not need to be watered as often as nonsucculent plants in containers. Once or twice a week should be plenty for most container-grown sedums unless the temperatures are extremely hot or the pot is quite small.

Sometimes potted displays benefit from a feeding, particularly if they've been growing in the same container for a while and any nutrition has long since leached out of their mix (but, in that case, you would do better to repot the plants).

Soil mixes become depleted and compacted over time. Plus, the plant may have grown too big for its container. Divide it into two or more pieces, or move it into a larger pot. This is a handle-with-care project, as the foliage of some sedums is brittle and roots can be scraggly. The ideal time to do this is spring, when growth is surging. The plants are stronger and more energetic then, thus more likely to take to a new home quickly. Never repot or divide when a plant is in full bloom—it's stressful.

Cut Flowers

While any sedum's flower stalk could be snipped off the plant and brought indoors to a vase, the stonecrop types are often too small. The border sedums (hylotelephiums), on the other hand, are fine. Pick the flowerheads while still green (unopened-buds stage) to add accent to mixed bouquets. Or pick them when they color up in autumn and create an arrangement of compatible hues; cut just as they are beginning to open.

As you've observed, the border sedum flowerheads dry out well in the garden. If you like the late-season colors of russet, tan, and chocolate brown, harvest them then and use them in dried arrangements. At this stage, they are durable and will not shed seeds or petals all over your tablecloth. One of my favorite combinations in the garden as well as in a vase in the house is a few of these with black-eyed Susan and *Miscanthus sinensis* 'Morning Light'.

Pests and Diseases

While sedums tend to be trouble-free, you may encounter problems. As a nurseryman, believe me, I have seen them all! If your plants develop issues, there is an explanation and, in some cases, a remedy. Bear in mind that if you buy healthy plants from a reputable supplier and follow my advice regarding siting, planting, and care, you should be fortunate not to experience any problems. Healthy sedums are likely to remain so.

PESTS

Most of the time, sedums are not bothered by pests. Natural predators such as ladybugs, lacewings, parasitic wasps, and chickadees usually keep things under control. When problems do arise, intervention is important because pests can be vectors for fungal diseases and viruses. Watch over your plants. If something attacks them, make sure you notice sooner rather than later. Quarantine new plants and afflicted ones until you are sure they are clear.

Leaf damage from an insect.

If you decide to spray, check that the pest you are after is listed on the label. Then, always follow the product directions regarding timing, amount, and application method. Many insecticides can damage succulent foliage, so weigh your options before treating an entire plant or patch. In particular, avoid malathion. Severely infested, badly damaged plants ought to be discarded to avoid spread.

At my nursery, aphids are the most common pest. These tiny insects relish new, tender shoots and are most problematic in our cool greenhouses when many of the natural predators are not active. Containerized sedums brought inside a home for winter are vulnerable for the same reason. For control, we use horticultural oil, neem oil, or insecticidal soaps at a low dosage, applying a solution every two to three days for a week or two to keep the aphid population from exploding. We avoid spraying in full sun because it can scorch foliage. For whatever reason, favorite targets of aphids are the border sedums *Hylotelephium cauticola*, *H. erythrostictum*, and *H. sieboldii*.

Mealybugs are most common on the roots or underground stems of stressed or overfertilized plants, where they suck out nutrients. Potted sedums are a favorite target. These pink-bodied insects hide in a white woolly mass, but if you pop a rootball out of a pot, the masses are easily viewed at the edge of the soil. For a small infestation, brush off the bugs, wipe them off with a cotton ball dipped in rubbing alcohol, or clean up the rootball and repot in new mix. For a bigger problem, try one of the sprays recommended for aphids or a pyrethrum-based insecticide.

Slugs and snails are known to nibble the foliage of thick-leaved sedums like *Hylotelephium sieboldii* and its cultivars. Damp and humid conditions favor the appearance of these mollusks, so avoid overwatering plants. Spread slug baits around the base of plants to protect them. If the problem persists, reapply regularly and especially after rain. Where possible, take steps to dry out an affected area.

Black wine weevil and strawberry root weevil are most common in milder climates, and a big pest on some agricultural crops, but they have not been a problem at my nursery. The grubs (little white worms with a black head) feed on the roots, while the adults feed on the edge of the foliage. When buying potted plants, lightly tug the foliage and if the plant pops right out, showing little root development, it could be a sign that it has weevils or other health problems. Your best defense is not to buy afflicted plants in the first place. If weevils do appear on your sedums, discard the plants and replace the soil in that area before planting anything else.

Root knot nematodes are another possible pest. These soil-borne microscopic worms feed on roots during warm summer months, causing galls to form on the roots and

Deer-Proof?

SEDUMS APPEAR IN deer-resistant plant lists from time to time. Still, there is no guarantee that deer won't eat your sedums, particularly if they don't have better options or if it's a long winter. You can try deterring the marauders with sprays or tall fencing. Should they manage to nibble your sedums, though, oftentimes the plants survive, thanks to their tenacious root systems and natural resilience.

damaging a plant's ability to absorb water and nutrients. The only way to make a firm identification is to submit soil and root samples to a lab; however, affected plants lose vigor and will wilt even where there is enough water. To minimize damage, plant sedums during cool times of the year when the nematodes are not feeding, that is, when temperatures are below 64°F (18°C).

DISEASES

Much more problematic to sedums than pests is their susceptibility to diseases. Before I get into the details, remember the importance of prevention. Keep after weeds, as they can harbor pests that spread diseases. Clean up dead plant tissue promptly, which can also harbor or encourage disease. Unless a variety has evergreen foliage or persistent seedheads that stand up all winter, I would recommend cutting back your sedums in late fall.

If a plant gets really sick and debilitated, just take it out and get rid of it. In the nursery, we always pull out and destroy those that show symptoms or signs of these diseases since problems can spread rapidly via irrigation. Another thing we have started doing is raising some of the susceptible varieties in biodegradable rice pots, which dry out faster, keeping the plants healthier.

Established, older sedums of any kind appear to be more resistant. In other words, the younger the plant, the more vulnerable it is. Hylotelephiums are particularly prone to fungal diseases, except for some of the new hybrids that have *Hylotelephium spectabile* as a parent. In the nursery, anyway, new *H. telephium* cultivars and hybrids are vulnerable.

Rhizoctonia, which causes root root, and *Phythium*, which causes collar rot or damping off, typically occur in the heat of summer when moisture levels are high. Pale green foliage, wilting, and stunted growth are warning signs. Prevention is the best way to minimize damage from these two soil-borne fungi: allow more space between plants and reduce water. A biological control called *Trichoderma harzianum* works preventatively to control root rot.

Another fungal disease is caused by *Phoma telephii*. Symptoms include sunken spots on foliage or stems. Damage is especially noticeable on red-leaved hylotelephiums, which seem to be more prone to leaf issues since they can be more easily burned by late frosts and scorch when the foliage is wet and the sun is strong. The red-leaved hylotelephiums may even be slightly less hardy because of this. Damp soil, damp foliage, and overwatering encourage this growth of this fungus, so do whatever you can to dry out and aerate the growing area. Cut out and dispose of affected stems and leaves.

Sclerotium rolfsii causes rot to stems and roots. It can appear as small yellow dots known as sclerotia at the base of the plant. Warm temperatures, usually 80°F (57°C), soil that is low in nitrogen, and wet weather encourage the affected areas to develop a white cottonlike growth. Infected plants look wilted, with the stems collapsing or being girdled at the soil line. Once infected, plants should be destroyed and the soil should be removed 6 inches (15 cm) beyond.

The broad flowerheads of a border sedum contribute contrasting color and texture in this summer planting.

Botrytis cinerea, sometimes called gray mold, grows a hairy, gray mass. It too is caused by excessive moisture on the leaves in cooler temps of 50–70°F (10–21°C). Afflicted leaves get irregular-shaped brown spots, usually surrounding a vein. Border sedums are susceptible to this fungus, most of the ground-hugging stonecrops are not. The best prevention is good air movement. Also, keep the foliage off the soil, if possible. Mulching helps block the fungus from the plants.

Fungal leaf spot, or anthracnose, appears as a white mass around the lower stems and sometimes on the foliage. Sunken spots develop on the leaves, making them very unattractive. Because wet conditions encourage this fungus to grow, water plants in the morning only, and try to avoid splashing the foliage. Do whatever you can to dry out and aerate the growing area. When the plant is dry, pick off and dispose of infected leaves. In fall, cut the plants down and dispose of them (but do not add diseased plants to your compost pile).

Powdery mildew (*Erysiphe*) is mainly a problem for greenhouse-grown stonecrops. The disease thrives when humidity is high, temperature is between 68 and 86°F (20–30°C), and light levels are low. Symptoms are small spots of white powder on the tops of leaves. While it does not kill plants, it does mar their appearance. To prevent powdery mildew on plants, space plants sufficiently apart to allow air circulation, keep humidity low, and ventilate the greenhouse. Remove any afflicated parts and let the plants rejuvenate. If a plant looks really bad, you should tear it out and replace it.

VIRUSES

While odds are small that viruses will bother your sedums, the fact is, sedums are susceptible to certain of them. I got to learn of plant viruses first-hand when one of my plants was tested and the results came back revealing it had three different viruses: cucumber mosaic virus (CMV), potato virus Y (POTY), and tobacco ring spot virus (TRSV).

Plant viruses are spread by a vector, typically a sucking insect or nematode, but can be passed physically plant to plant by sap. This physical transfer can happen when animals feed on both an infected plant and then a clean plant, or during the process of pruning. In the case of TMV, the virus can even be introduced from a smoker touching the foliage of a plant. Certain plants are more susceptible, but I would not recommend smoking and working with plants.

The physical sign of a virus is usually a chlorosis or spotting of foliage and sometimes a puckering of foliage. The main issue with having a plant with a virus is loss of vigor which will allow other diseases to attack the plant and eventually kill it.

OTHER PROBLEMS

Edema can be a debilitating problem for sedums. This physiological disorder, named for a word that means "swelling," develops on plants that absorb water faster than they are losing it. It can happen when the weather is humid and very wet or you overwater; it also afflicts sedums in moist soil during times of cool, moist air.

A bee feeds on the nectar-rich flowers of *Hylotelephium* 'Pure Joy'.

When sedums get edema, leaves swell in irregular patches then burst, leaving tan to brown patches that resemble insect-feeding or fungal damage. Not only does this look terrible, but it weakens the plant, making it vulnerable to pests and diseases. Once leaves are affected, there is no cure.

I have mostly seen this occur on *Hylotelephium* hybrids like 'Matrona'. To prevent more damage, cut back on watering, water only in the morning, and do not let water splash on foliage. If possible, move your plants to a spot with more light and air movement. Alternatively, cut them back hard, and hope that when new foliage appears, conditions are drier and the plants will be healthy.

Hylotelephium 'Black Jack' reverting to 'Matrona' and also displaying edema.

A final concern when growing sedums is weeds, particularly when it comes to establishing and maintaining sedums. Weeds can harbor pests, which tend to spread disease, so keeping weeds in check is an integral part of keeping plants healthy. In the case of a greenroof, maintenance is essential until the entire roof is covered in the desired vegetation.

Propagation

For the most part, making more sedums is easy. You may use division, cuttings, or even seed. Division will make the most sense for home gardeners. Many times, a plant will begin to root even before it is separated from the mother plant. At the nursery, where we aim to produce sedums in quantity, we usually use stem cuttings—it's simple enough to do this at home, on a smaller scale. Both division and cuttings lead to plants that are identical to their parent plant.

It is also possible to raise sedums from seed. If you use seed from a species, the babies will be like the parent plant. If from a hybrid, or if you are uncertain, the results may be disappointing since some hybrid plants are sterile.

DIVISION

This is the best method for home gardeners. It's easy and successful. Work with mature or larger plants.

Best time: in early spring when new green shoots are visible.

Recommended for: the large-rooted *Hylotelephium* and *Rhodiola*.

Equipment/supplies needed: a trowel for digging and replanting; a clean, sharp knife for cutting.

Start with: a rooted plant. Begin at the center and slice toward the outer edge. I recommend halving or quartering. Make sure each piece includes green buds; ideally you will want a multiple eye or stemmed division, with multiple roots.

How to proceed: plant the divisions to the same soil depth as before, water, and tend them with care over the next few weeks until they become established.

Sedums are easily raised from stem cuttings in flats of a well-drained soil mix.

Some sedums make air roots, which makes taking viable cuttings easy.

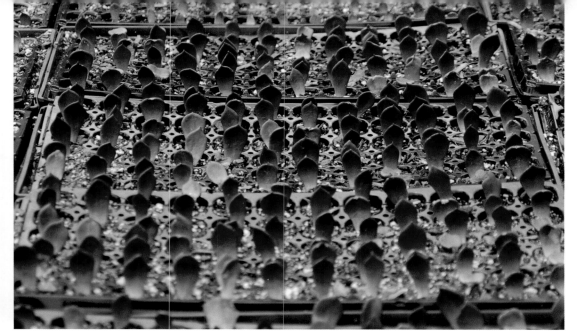

Leaf cutting propagation at a nursery. Each leaf should have a bud at the base if possible; lightly press the base into a sterile growing medium.

STEM CUTTINGS

With this technique you can make quite a few new plants from just one plant.

Best time: Anytime the plant is not budding or flowering is fine; the procedure is trickier in very hot weather in midsummer. Mostly we do the small species like *Sedum acre*, *S. album*, and *S. hispanicum* cultivars before it gets too hot, and move on to the larger species and deciduous species in midspring through early summer. Moving in to fall some species begin to bud so this is not their ideal time.

Recommended for: Any kind of sedum, whether the bigger *Hylotelephium* types or the lower-growing stonecrop types. Of the latter, *Sedum album* and *Sedum dasyphyllum* are particularly easy.

Equipment/supplies needed: Trays or pots 2–3 inches (5–7.5 cm) deep, well-drained potting soil, and a sharp, clean knife for cutting, if necessary. For watering, you can either set a tray of stem cuttings in a larger tray to soak up some moisture, or mist from above.

Start with: One healthy stem. Cutting size varies, depending on the species and the time of year. Generally you want one that is 1–2 inches (2.5–5 cm) long, maybe 3 inches (7.5 cm) if you are working with a larger-leaved plant. Each section should have a few leaves at least. For trailing types, gently strip leaves off the stem.

How to proceed: Lightly push the base of a stem into a pot or tray of well-drained potting soil—this is called "sticking the cutting." Water carefully at least once a day, maybe twice. But do not let their medium become constantly wet; it is better to let cuttings dry out a little bit between waterings. Rooting can take two to three weeks, and as the plants become rooted they can be watered less.

SEDUMS FROM SEEDS

In nature, sedums are insect pollinated. So it comes as no surprise that they are extra-tiny and lightweight. Most are at most ⅛ inch (3 mm) long, and very thin. To raise new plants from them, you must handle them with care.

It is possible to buy seed or you can get some from another gardener or a club (such as a rock-garden or succulent group's seed exchange). I collect seed from my own plants, too. Here's what to do: Cut off a seedhead, when flowering is finished and as it begins to turn from green to brown. Place this in a paper bag and store in a cool, dry place for two or more weeks to dry. Then take it out and thresh it over a very fine screen, so the tiny seeds can fall through and the debris remains on the screen. Carefully put the seeds into a plastic bag, and don't forget to label them with the plant's name and the date. Seeds may be stored in a refrigerator for up to a year. You may sow seed harvested from fall-bloomers the following spring.

Best time: Spring, when temperatures are between 40 and 70°F (4–21°C).

Recommended for: any species types (not for cultivars, which should be propagated vegetatively).

Equipment/supplies needed: Flats or pots; fine-textured, sterile seed-starting soil mix.

Start with: Ripe seed.

How to proceed: Gently press seeds into pre-dampened mix (do not cover them with mix). If using flats, sow them in rows. Place the container where it will not be exposed to wind or drafts that might dislodge the seeds. Keep the soil moist. Germination usually takes two to three weeks.

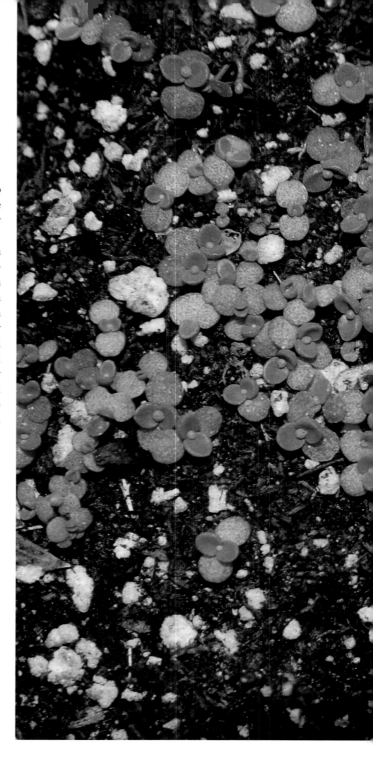

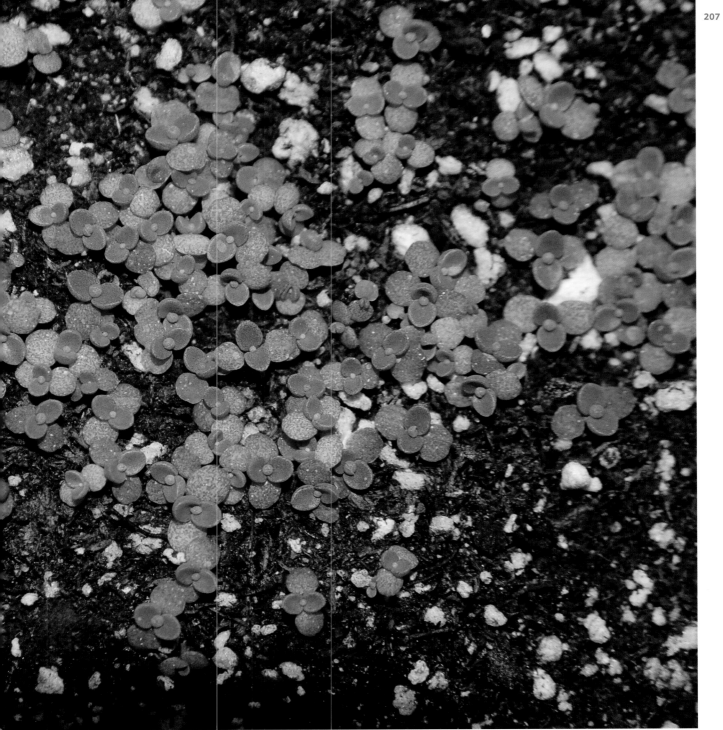

A tray of newly germinated sedum seedlings.

WHERE TO BUY

AUSTRALIA

The Succulent Garden
20 Pinus Avenue
Glenorie, NSW 2157
www.thesucculentgarden.com.au

CANADA

Pacific Rim Native Plant Nursery
P.O. Box 413
Chilliwack, British Columbia V2P 6J7
www.hillkeep.ca

UNITED KINGDOM

Alpine and Grass Nursery
Northgade
West Pinchbeck
Spalding
Lincolnshire
England PE11 3TB
www.alpinesandgrasses.co.uk

Barnsdale Gardens
The Avenue
Exton
Oakham
Rutland
England LE15 8AH
www.barnsdalegardens.co.uk

Beth Chatto Gardens
Elmstead Market
Colchester
Essex
England CO7 7DB
www.bethchatto.co.uk

Burncoose Nurseries
Gwennap
Redruth
Cornwall
England TR16 6BJ
www.burncoose.co.uk

Cally Gardens
Gatehouse of Fleet
Castle Douglas
Kirkcudbrightshire
England DG7 2DJ
www.callygardens.co.uk

Cotswold Garden Flowers
Sands Lane
Badsey, Evesham
Worcestershire
England WR11 7EZ
www.cgf.net

Inshriach Nursery
Aviemore
Inverness-shire PH22 1QS
www.inshriachnursery.co.uk

Knoll Gardens
Hampreston
Wimborne
England BH21 7ND
www.knollgardens.co.uk

Marchants Hardy Plants
2 Marchants Cottages
Mill Lane
Laughton
East Sussex
England BN8 6AJ
www.marchantshardyplants.co.uk
No mail order.

Sedum Supply
Severn Business Centre
15 Severn Farm Enterprise Park
Welshpool
England SY21 7DF
www.sedumsupply.co.uk
Green roof plant specialist.

Slack Top Nurseries
1 Waterloo House
Slack Top
Hebden Bridge
England HX7 7HA
www.slacktopnurseries.co.uk

UNITED STATES

Alplains
P.O. Box 489
Kiowa, Colorado 80117
www.alplains.com
Mail-order seed only.

American Meadows
223 Avenue D, Suite 30
Williston, Vermont 05495
www.americanmeadows.com

Annie's Annuals and Perennials
801 Chesley Avenue
Richmond, California 94801
www.anniesannuals.com

Arrowhead Alpines
1310 North Gregory Road
P.O. Box 857
Fowlerville, Michigan 48836
www.arrowheadalpines.com

Avant Gardens
710 High Hill Road
Dartmouth, Massachusetts 02747
www.avantgardensne.com

Busse Gardens
17160 245th Avenue
Big Lake, Minnesota 55309
www.bussegardens.com

Canyon's Edge Plants
11691 West Country Club Road
Canyon, Texas 79015
www.canyonsedgeplants.com

Cistus Nursery
22711 NW Gillihan Road
Portland, Oregon 97231
www.cistus.com

The Flower Factory
4062 County Road A
Stoughton, Wisconsin 53589
www.theflowerfactorynursery.com

Garden Crossings
4902 96th Avenue
Zeeland, Michigan 49464
www.gardencrossings.com

Great Garden Plants
P.O. Box 1511
Holland, Michigan 49424
www.greatgardenplants.com
Mail order only.

Gulley Greenhouse and Garden Center
6029 South Shields
Fort Collins, Colorado 80526
www.gulleygreenhouse.com

High Country Gardens
P.O. Box 22398
Santa Fe, New Mexico 87502
www.highcountrygardens.com
Mail order only.

Jelitto Perennial Seeds
125 Chenoweth Lane, Suite 301
Louisville, Kentucky 40207
www.jelitto.com
Mail order seed only.

Jost Greenhouses
12348 Ecklemann Lane
St Louis, Missouri 63131
www.jostgreenhouses.com

Joy Creek Nursery
20300 NW Watson Road
Scappoose, Oregon 97056
www.joycreek.com

J. W. Jung Seed Company
335 S. High Street
Randolph, Wisconsin 53956
www.jungseed.com

Klehm's Song Sparrow Perennial Farm
13101 E. Rye Road
Avalon, WI 53505
www.songsparrow.com
Mail order only.

Laporte Avenue Nursery
1950 Laporte Avenue
Fort Collins, Colorado 80521
www.laporteavenuenursery.com

Lazy S'S Farm Nursery
2360 Spotswood Trail
Barboursville, Virginia 22923
www.lazyssfarm.com

Mary's Plant Farm
2410 Lanes Mill Road
Hamilton, Ohio 45013
www.marysplantfarm.com

Mesa Garden
P.O. Box 72
Belen, New Mexico 87002
www.mesagarden.com

Millcreek Gardens
3500 South 900 East
Salt Lake City, Utah 84106
www.millcreekgardens.com

Mountain Crest Gardens
P.O. Box 1023
Fort Jones, California 96032
www.mountaincrestgardens.com
Mail order only.

Plant Delights Nursery
9241 Sauls Road
Raleigh, North Carolina 27603
www.plantdelights.com

Simply Succulents Nursery
43785 Highway 63
Cable, Wisconsin 54821
www.simplysucculents.com

SMG Succulents
P.O. Box 148
Eagle Creek, Oregon 97022
www.smgsucculents.com
Mail order only.

Spring Hill Nurseries
110 West Elm Street
Tipp City, Ohio 45371
www.springhillnursery.com

Walnut Hill Greenhouse
217 Wheeler Road
Litchfield, Connecticut 06759
www.walnuthillgreenhouse.com
Mail order only.

White Flower Farm
P.O. Box 50, Route 63
Litchfield, Connecticut 06759
www.whiteflowerfarm.com

Wild Ginger Farm
24000 S. Schuebel School Road
Beavercreek, Oregon 97004
www.wildgingerfarm.com

Wyoming Plant Company
358 South Ash Street
Casper, Wyoming 82601
www.wyomingplantcompany.com

Yucca Do Nursery
P.O. Box 1039
Giddings, Texas 78942
www.yuccado.com

WHERE TO SEE

Devonian Botanic Garden
University of Alberta
Highway 60
Edmonton, Alberta T6G 2E1
www.devonian.ualberta.ca

Montreal Botanical Garden
4101, rue Sherbrooke Est
Montréal, Québec H1X 2B2
www.espacepourlavie.ca/en/
botanical-garden

**Nova Scotia Agricultural College
Rock Garden**
Dalhousie University
Halifax, Nova Scotia B3H 4R2
www.dal.ca/about-dal/agricultural-cam-
pus/about/gardens/rock-garden.html

Rock Wall Gardens
995 Code Road
Perth, Ontario K7H 3C8
www.rockwallgardens.com

Toronto Botanical Garden
777 Lawrence Avenue East
Toronto, Ontario M3C 1P2
www.torontobotanicalgarden.ca

**University of British Columbia
Botanical Garden and Centre for
Plant Research**
6804 SW Marine Drive
Vancouver, British Columbia V6T 1Z4
www.botanicalgarden.ubc.ca

The Alpine Garden Society
AGS Centre
Avon Bank
Pershore
Worcestershire
England WR10 3JP
www.alpinegardensociety.net

**National Collection of Sedum/Sedum
Society**
Mr. and Mrs. Ray Stephenson
8 Percy Gardens
Choppington
Northumberland
England NE62 5YH
www.cactus-mail.com/sedum

Phoenix Perennial Plants
Paice Lane, Medstead
Alton
Hampshire
England GU34 5PR
+44 (0) 1420 560695

Scottish Rock Garden Society
P.O. Box 14063
Edinburgh EH10 YE
Scotland
www.srgc.net

UNITED STATES

Allen Centennial Gardens
620 Babcock Drive
Madison, Wisconsin 53706
www.allencentennialgardens.org

Atlanta Botanic Garden
1345 Piedmont Avenue NE
Atlanta, Georgia 30309
www.atlantabotanicalgarden.org

Betty Ford Alpine Gardens
500 South Frontage Road
Vail, Colorado 81657
www.bettyfordalpinegardens.org

Boerner Botanic Gardens
9400 Boerner Drive
Hales Corners, Wisconsin 53130
www.boernerbotanicalgardens.org

Brooklyn Botanic Garden
1000 Washington Avenue
Brooklyn, New York 11225
www.bbg.org

Cheekwood Botanical Garden
1200 Forrest Park Drive
Nashville, Tennessee 37205
www.cheekwood.org

Chicago Botanic Garden
1000 Lake Cook Road
Glencoe, Illinois 60022
www.chicagobotanic.org

Denver Botanic Gardens
1007 York Street
Denver, Colorado 80206
www.botanicgardens.org

Huntington Botanical Gardens
1151 Oxford Road
San Marino, California 91108
www.huntington.org

Idaho Botanical Garden
2355 Old Penitentiary Road
Boise, Idaho 83712
www.idahobotanicalgarden.org

Lady Bird Johnson Wildflower Center
4801 La Crosse Avenue
Austin, Texas 78739
www.wildflower.org

Lake Harriet Peace Garden
4125 E. Lake Harriet Parkway
Minneapolis, Minnesota 55409
www.minneapolisparks.org/default.
asp?PageID=4&parkid=349

Longwood Gardens
1001 Longwood Road
Kennett Square, Pennsylvania 19348
www.longwoodgardens.org

Memphis Botanic Garden
750 Cherry Road
Memphis, Tennessee 38117
www.memphisbotanicgarden.com

Minnesota Landscape Arboretum
3675 Arboretum Drive
Chanhassen, Minnesota 55318
www.arboretum.umn.edu

Missouri Botanical Garden
4344 Shaw Boulevard
St. Louis, Missouri 63110
www.missouribotanicalgarden.org

New York Botanical Garden
2900 Southern Boulevard
Bronx, New York 10458
www.nybg.org

Olbrich Botanical Gardens
3330 Atwood Avenue
Madison, Wisconsin 53704
www.olbrich.org

Oregon Garden
879 West Main Street
Silverton, Oregon 97381
www.oregongarden.org

Rotary Botanical Gardens
1455 Palmer Drive
Janesville, Wisconsin 53545
www.rotarybotanicalgardens.org

Tizer Botanic Gardens
38 Tizer Lake Road
Jefferson City, Montana 59638
www.tizergardens.com

**University of California
Botanical Garden at Berkeley**
200 Centennial Drive
Berkeley, California 94720
www.botanicalgarden.berkeley.edu

FOR MORE INFORMATION

BOOKS

Baldwin, Debra Lee. 2007. *Designing With Succulents*. Portland, Oregon: Timber Press.

Cave, Yvonne. 2004. *Succulents for the Contemporary Garden*. Portland, Oregon: Timber Press.

Clausen, Robert Theodore. 1911. *Sedums of North America, North of the Mexican Plateau*.

Dunnett, Nigel and Kingsbury, Noel. 2004. *Planting Greenroofs and Living Walls*. Portland, Oregon: Timber Press.

Evans, Ronald L. 1983. *Handbook of Cultivated Sedums*. Middlesex, United Kingdom: Science Reviews Ltd.

Froderstrom, H. A. 1936. *The Genus Sedum L*. Goteborg

Payne, Helen E. 1972. *Plant Jewels of the High Country*. Medford, Oregon: Pine Cone Publishers.

Praeger, Lloyd Robert. 1967. *An account of the Genus Sedum as found in Cultivation*. New York: Stechert-Hafner Service Agency, Inc.

Snodgrass, Edmond C. and Snodgrass, Lucie L. 2006. *Greenroof Plants*. Portland, Oregon: Timber Press.

Stephenson, Ray. 1994. *Sedum: Cultivated Stonecrops*. Portland, Oregon: Timber Press.

Whitehouse, Christopher. 2007. *Herbaceous Sedums*. Surrey, United Kingdom: Royal Horticultural Society.

WEBSITES

All Things Plants database: www. allthingsplants.com

Avani Plants: www.avaniplants.com

Drought Smart Plants: www. drought-smart-plants.com

Future Plants: www.futureplants.com

GardenWebcommunity: www.gardenweb.com

Great Plant Picks: www.greatplantpicks. org

Green Roofs: www.greenroofs.com

International Crassulaceae Network: www.crassulaceae.net

Wayne Fagerlund: www.sedumphotos.net

ORGANIZATIONS

Cactus and Succulent Society of America: www.cssainc.org

North American Rock Garden Society: www.nargs.org

Royal Horticultural Society: www.rhs.org.uk

Sedum Society: www.cactus-mall.com/sedum

HARDINESS ZONE TEMPERATURES

USDA ZONES & CORRESPONDING TEMPERATURES

Temp °F			Zone	Temp°C		
−60	to	−55	1a	−51	to	−48
−55	to	−50	1b	−48	to	−46
−50	to	−45	2a	−46	to	−43
−45	to	−40	2b	−43	to	−40
−40	to	−35	3a	−40	to	−37
−35	to	−30	3b	−37	to	−34
−30	to	−25	4a	−34	to	−32
−25	to	−20	4b	−32	to	−29
−20	to	−15	5a	−29	to	−26
−15	to	−10	5b	−26	to	−23
−10	to	−5	6a	−23	to	−21
−5	to	0	6b	−21	to	−18
0	to	5	7a	−18	to	−15
5	to	10	7b	−15	to	−12
10	to	15	8a	−12	to	−9
15	to	20	8b	−9	to	−7
20	to	25	9a	−7	to	−4
25	to	30	9b	−4	to	−1
30	to	35	10a	−1	to	2
35	to	40	10b	2	to	4
40	to	45	11a	4	to	7
45	to	50	11b	7	to	10
50	to	55	12a	10	to	13
55	to	60	12b	13	to	16
60	to	65	13a	16	to	18
65	to	70	13b	18	to	21

FIND HARDINESS MAPS ON THE INTERNET.
United States *http://www.usna.usda.gov/Hardzone/ushzmap.html*
Canada *http://www.planthardiness.gc.ca/* or *http://atlas.nrcan.gc.ca/site/ english/maps/environment/forest/forestcanada/planthardi*
Europe *http://www.gardenweb.com/zones/europe/* or *http://www.uk. gardenweb.com/forums/zones/hze.html*

ACKNOWLEDGMENTS

Without God, there would be no sedums. Thanks to God for these beautiful plants.

There are many people to acknowledge in the making of this book. First I'd like to acknowledge Ray Stephenson for his book on sedums, all his help, and the work he does to promote and educate people about these great plants through the Sedum Society.

I'd also like to thank my wife Therese and my family for supporting me in my business, my brother Kurt, my sister Sonja, and my mom Trudy. The nursery is where I met my wife and where I got to work side-by-side with my dad, one of my mentors.

Others to thank include: Darrel Probst, Wayne Fagerlund, Dave McKenzie, Ed Snodgrass, Galen Gates, Karl Batschke, Margit Bischofberger, Lauren Springer Ogden, John Greenlee, Paul Little, Pierre Bennerup, Marc Laviana, Harlan Hamernick, Chris Hansen, Roy Diblik, Roy Klehm, George Radtke, Panayoti Kelaidis, Ozzie Johnson, and Barry Yinger, Lois Hoveke, Renee Jaeger, William Rupert, Dagmar Petrlíková. Special thanks to Maria Kelly for typing. Extra thanks to Teri Dunn Chase for her finesse with words in the editing and writing process. The book is more enjoyable thanks to her.

I owe a great debt to fellow sedophiles, especially the members of the Sedum Society. That group's yearly cutting exchange gave me the ability to meet and grow many new plants; its quarterly newsletter also educated me and fueled my passion for the ever-expanding world of sedums. Through it all, I realized that I wasn't the only one trying to find my way. In 2012, *Sedum* was listed as the fifth most-searched plant on the Better Homes and Gardens website, suggesting that ordinary gardeners, not just specialists or hobbyists, are looking for these plants and seeking information about them. So I leapt at the opportunity to set down what I have learned about sedums in this book.

PHOTO CREDITS

Photographs are by the author unless indicated otherwise.

COVER: (front) iStockphoto/zorani; (back bottom right) Chris Hansen.

WAYNE FAGERLUND/SEDUMPHOTOS.NET, pages 98, 99, 149, and 184 left.

CHRIS HANSEN, page 70 right.

TERESA HORVATH, page 228.

DAGMAR PETRLÍKOVÁ, pages 65, 106, 116, 138 right, 144, 176, and 182.

SCOTT SELLERS/BRT AERIALS, page 36 bottom left.

TERRA NOVA NURSERIES, pages 70 left, and 97 right.

ISTOCKPHOTO/AIMINTANG, pages 188–189.

ISTOCKPHOTO/ZORANI, pages 4–5.

SHUTTERSTOCK/ANNAVEE, pages 48–49.

SHUTTERSTOCK/UGU, pages 62–63.

SHUTTERSTOCK/USBFCO, pages 10–11.

INDEX

ABOUT THE AUTHOR

BRENT HORVATH is a third-generation gardener, second-generation nurseryman, and first-generation American. When he was growing up, his parents had a garden center, a landscaping service, and a florist shop, so he was exposed to a wide range of plants and gardening. Among the family's field-dug offerings, the handsome and dependable *Sedum* 'Autumn Joy' (now called *Hylotelephium* 'Herbstfreude') was a staple. He also remembers the mats of dense-growing *S. spurium* (now called *Phedimus spurius*) right outside the front door at home.

After graduating from Oregon State University with a degree in ornamental horticulture, Brent returned to the Midwest and started growing finished perennials for the wholesale trade. Today, he is the proud—if sometimes exhausted—proprietor of Intrinsic Perennial Gardens in Hebron, Illinois. Although his nursery grows a range of plants, it has a strong focus on ornamental grasses and sedums, plants that are well-suited to today's wishes for naturalistic, low-maintenance landscaping. The nursery is increasingly serving the exciting trend of greenroof landscaping, for which sedums are ideal.

Brent's enthusiasm for sedums has also led him to select and create new varieties, among them *Hylotelephium* 'Lajos' (named in tribute to his late father), *H.* 'Thundercloud', and *Phedimus spurius* 'Red Rock'. He currently has twelve plant patents including four sedums with two more sedum patents pending. Visit Brent at intrinsicperennialgardens.com.

Front cover: *Sedum* cultivars
Spine: *Sedum oreganum*
Title page: *Sedum hispanicum* var. *hispanicum*
Contents page: *Sedum* cultivars

The Haseltine Building   6a Lonsdale Road
133 S.W. Second Avenue Suite 450   London NW6 6RD
Portland, Oregon 97204-3527

For details on other Timber Press books and to
sign up for our newsletters, please visit our websites,
timberpress.com and timberpress.co.uk.

Library of Congress Cataloging-in-Publication Data
Horvath, Brent.
  The plant lover's guide to sedums/Brent Horvath.—1st edition.
    pages cm
  Includes bibliographical references and index.
  ISBN 978-1-60469-392-8
  1. Sedum. I. Title.
  SB413.S43H67 2014
  583'.72—dc23

                                              2013034033

A catalog record for this book is also available from the British Library.

Book and cover design by Laken Wright
Layout and composition by Laura Shaw Design
Printed in China

THE **PLANT LOVER'S GUIDE** TO
SALVIAS

JOHN WHITTLESEY

THE **PLANT LOVER'S GUIDE** TO
DAHLIAS

ANDY VERNON

THE **PLANT LOVER'S GUIDE** TO
SNOWDROPS

NAOMI SLADE